blue + white retreats

blue + white retreats

text by

Kannan Chandran

photography by

Luca Invernizzi Tettoni

Don Siegel

MANAGING EDITOR
Melisa Teo

ASSISTANT EDITOR
Ng Wei Chian

DESIGNER
Chan Hui Yee

PRODUCTION MANAGER
Sin Kam Cheong

First published in 2004 by **Archipelago Press**
an imprint of **Editions Didier Millet**
121 Telok Ayer Street, #03-01, Singapore 068590
Tel: (65) 6324 9260 Fax: (65) 6324 9261
Website: **www.edmbooks.com**

An exclusive production by **The Turning Point**,
trading under the name **Turning Point Holdings Pte Ltd**,
for and in cooperation with Club Med Asie SA. Unless
otherwise stated, all photographs appearing in this book are
owned by, and used with the permission of Club Med Asie SA.

Printed in Singapore

ISBN: 981-4068-98-5 (hardback)
981-4155-45-4 (paperback)

ADDITIONAL PHOTO CREDITS
PAGES 100–101: Guido Alberto Rossi
PAGE 106 (LEFT AND RIGHT): Krimo Guermond, Marine Prodive, Kani, The Maldives

PAGE 1: The Andaman Sea meets the horizon at a beach on Phuket.

PREVIOUS PAGES: Whether in the Maldives (PAGE 2) or on Lindeman Island (PAGE 3), the sun-dappled waters are equally inviting.

RIGHT: Sunny days are common on Lindeman Island, the warmest island in the Whitsundays grouping,

FOLLOWING PAGES: The vast expanse of the Great Barrier Reef extends in all its green- and blue-hued glory.

Lindemandive

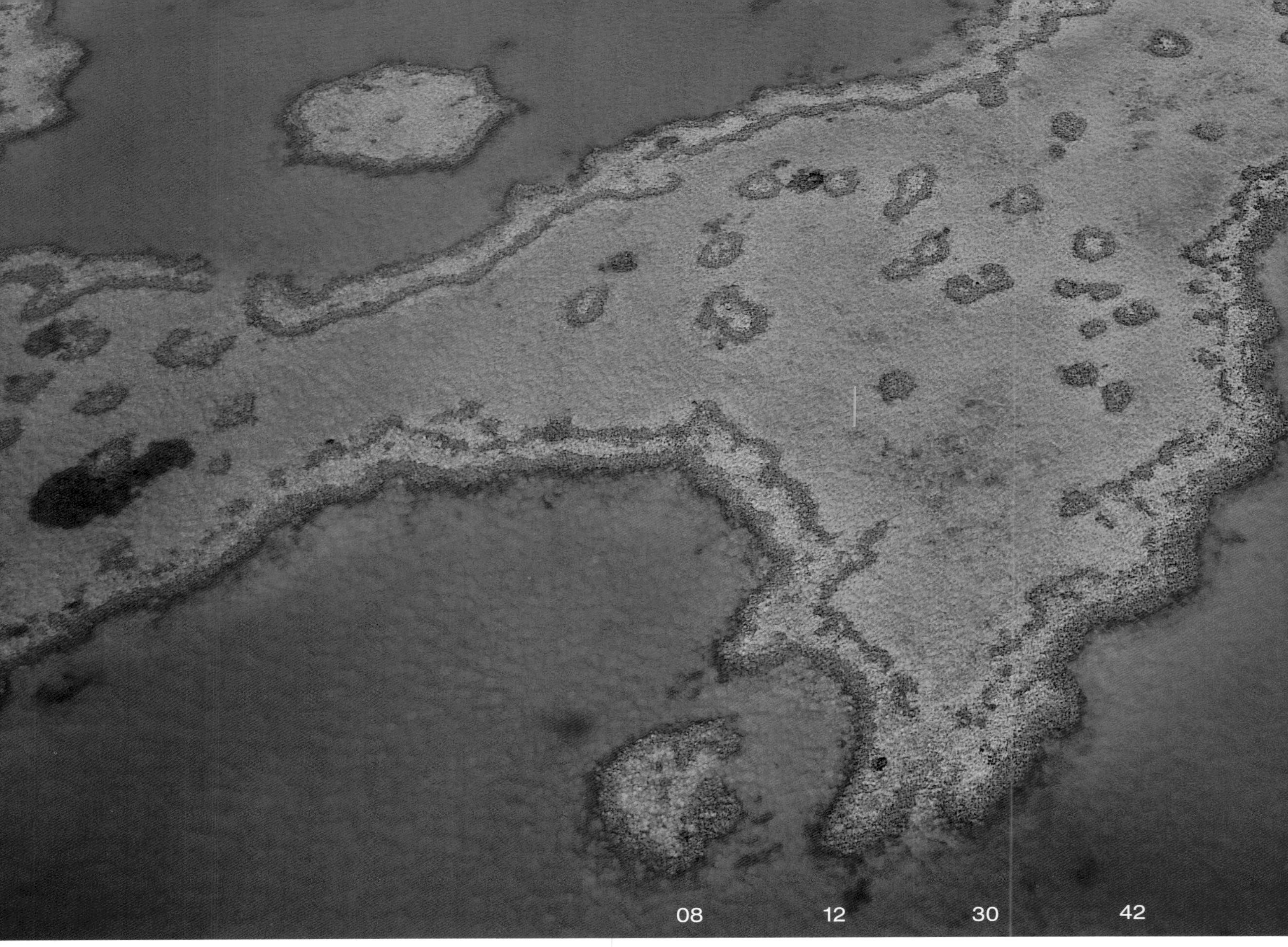

contents

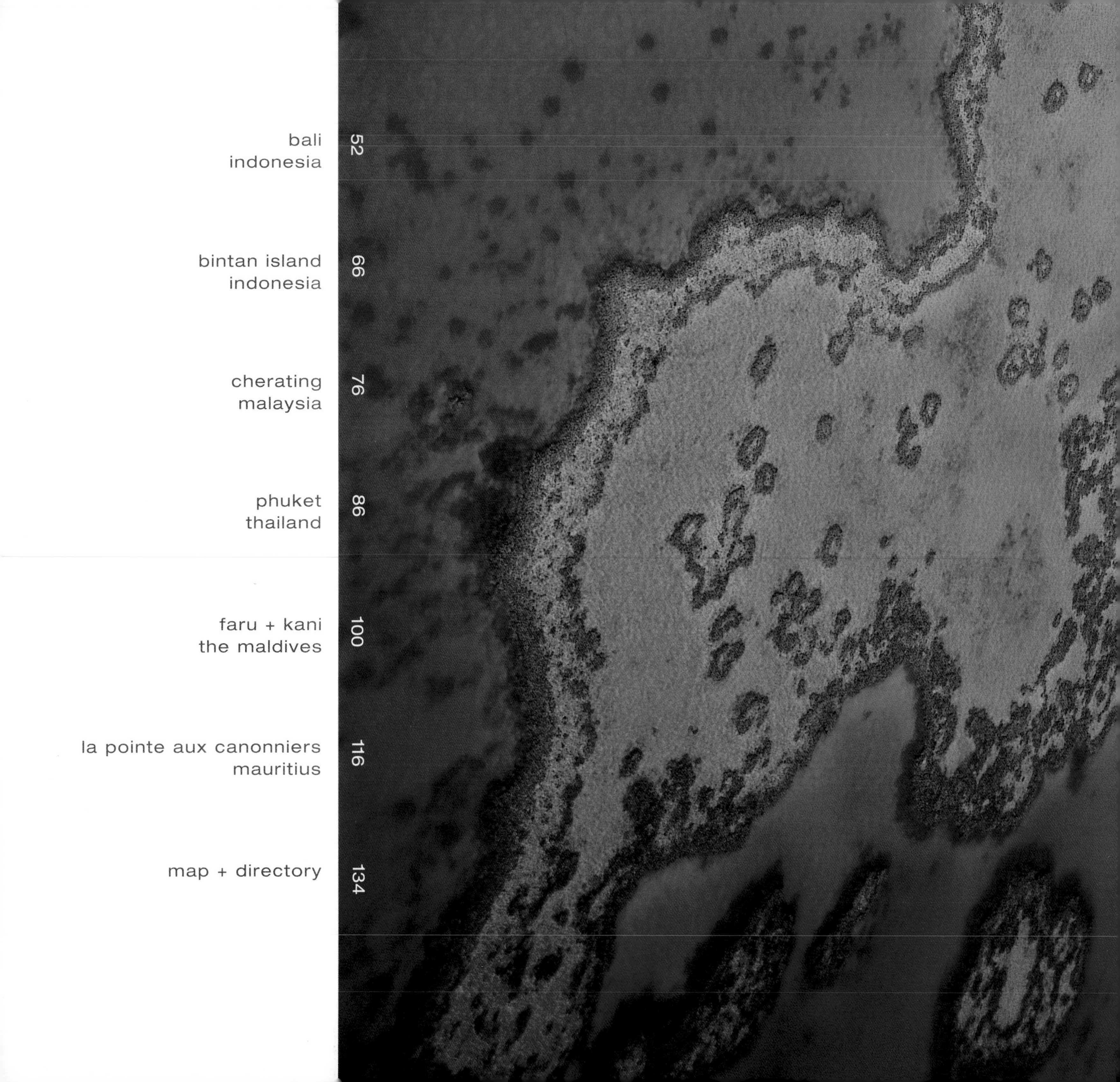

Club Méditerranée has always sourced the best locales for its facilities. It comes from years of experience in tirelessly scouting for the perfect spot, and from having set up villages all over the world, from Tahiti to the former Yugoslavia.

Club Med, as it is globally known these days, was the first to identify the potential of Bora Bora as a location for its resorts, and despite the remoteness of French Polynesia, it didn't deter the flood of tourists drawn both by its tropical charms and the Club Med way of life.

The Club Med philosophy has also benefited from a vision that has grown and moved with the times. Today, Club Med is for singles, couples, honeymooners, and families with their young

Colours conjure up strong images that are almost universal. The blue of the ocean and the sky meets the white of sand and clouds, painting ideal locations where nature's hues combine to create the perfect playground setting for Club Med's villages.

+ white

children. And whether you want to laze in the sun, or get caught up in the huge variety of activities available, you can have an incomparable holiday experience that will always prove memorable.

It was with all these in mind that *Blue and White Retreats* was conceived and created. From the idyllic turquoise waters of Bora Bora and the famed beaches of Bali to the atolls of the magical Maldives and secluded charms of Kabira in the Ryukyu Islands of Japan, the essence of Club Med's vitality and vigour are captured here in words and pictures. The variety that awaits the visitor to a Club Med location in the Asia-Pacific region is astounding, from the quiet beaches of Cherating in Malaysia and the bustle of Phuket in

Thailand, to the solitude of Kabira in Okinawa and the beauty of Lindeman Island off the Australian mainland. One can also discover the twin delights of Kani and Faru in the Maldives, the azure waters of Bora Bora, the beauty of Bali, the charm of Bintan and the natural magnificence of Mauritius.

The magic of Club Med is reflected not just in its locations, but in the passion with which the concept was realised. The concept of Club Med was born in 1950 on a deserted beach in the Balearic Islands of Spain, and initially took the shape of the tent village of Alcudia.

Gérard Blitz, a keen athlete and member of the French Resistance during World War II, founded Club Med on the basis of developing an "appreciation for the outdoor life and the practice of physical education and sports".

The first 300 guests arrived at Alcudia in June 1950, intrigued by the attractive brochures and posters, and happy for any diversion after the war. They found themselves thrilled at the friendliness of the place and the 'Club Spirit'.

Club Med introduced a whole new lifestyle which was centred around the beach, as well as a democratisation of scuba-diving and water-skiing, both sports which were traditionally limited to an elite crowd. The relaxed atmosphere was also credited with levelling social distinctions, bringing people from different classes together in their pursuit of the good life Club Med still prides itself on offering. To quote *Le Figaro*, each and every Club Med village was, and still is, a 'tiny democracy'.

This 'Spirit' has endured till today, right from the day a dedicated group of enterprising people built up the Club Med concept brick by brick into a business empire that has spelt fun and the celebration of the human spirit for adventure.

The G.O. (*gentil organisateur*, or gracious host) and G.M. (*gentil membres*, or gracious members) concepts for Club Med—organisers and guests respectively—also heralded the start of a novel relationship between guest and host, adding to and reinforcing the 'Club Spirit'.

The Club Med formula has unfailingly won praise at all its locations. The brilliant vistas provide a scenic backdrop to a wide programme of sporting activities and clubs catering to specific interests. Added to this is the non-stop party atmosphere that has long been a hallmark of Club Med and which G.M.s eagerly seek out.

Whether one is in search of solitude, peace and relaxation or an activity-filled fortnight, Club Med delivers all these and more. Novelty and spontaneity are the watchwords, with dynamism the key factor in the growth and success of Club Med. At Club Med, choices on offer are always expanding. We hope that by bringing you vivid images from Club Med's Asia-Pacific villages, the fun can last that much longer.

PAGE 8: Not to be missed is the talcum powder-white sand of Whitehaven Beach, near Lindeman Island. Made up of nearly pure silica, your feet sink into the soft sand as you leap off the chopper that brings you to this beach paradise.

OPPOSITE: Lush, giant Gorgonian coral provides a colourful welcome for divers in the Maldives.

TOP: Lazy days spent soaking up the sun are characteristic of a holiday spent at any Club Med village worldwide.

bora bora
french polynesia

When French Polynesia heaved into existence as the earth convulsed and then cooled in a colossal display of geographic splendour that went largely unnoticed, it was the creation of a magnificent, fitting stage for an epic drama that was to incorporate all the very best elements of a hit production—beauty, violence, conflict, and intrigue.

An explosive introduction

Names such as Tahiti and Bora Bora inevitably conjure up the usual images of tropical bliss: blue waters, lush vegetation, and coconut trees bending to seek out the sun.

If all this reminds you of paradise, you're not too far wrong. These islands have also inspired the likes of painter Paul Gauguin and various celluloid interpretations of the mutiny on the *HMS Bounty*.

Till today, no one is certain why the ancestors of the native Polynesians made the journey over such rough waters, presumably from Indonesia and the Philippines. When they arrived, they established a hierarchical community with its own distinct practices that included fishing as a means of subsistence, tattooing and cannibalism, among other seemingly unusual habits.

But their course of life was to change dramatically with the arrival of a succession of

Bora Bora, French Polynesia
Pacific Ocean
Latitude 15° S, longitude 144° W

European ships carrying curious crew members who sought to claim the land for themselves.

Provocative visitors

From 1500, the Spanish galleons were the first to stake their rights on the islands, followed by the Dutch, and then the British, led in 1765 by John Byron, grandfather of the poet Lord Byron. A year later, Samuel Wallis on the *HMS Dolphin* was the first European to visit Tahiti.

While Wallis was heading back with stories of 'The Island of Lust' and willing natives ready to offer themselves in bed for a nail (which would be used to make hooks for the fishing community), the great French Pacific explorer, Louis-Antoine de Bougainville, came into port, only to be succeeded by James Cook, bearing the Union Jack, and the Spanish, who were seeking to renew their original claim to the islands.

The biggest impact of this foreign influx, however, was to come in the shape of men of the cloth.The Protestant missionaries converted the local population, which consequently led to the fading of a number of local cultural practices and reformed the religious landscape.

Among all these many dizzying claims for ownership, perhaps the most lasting impression is left by the story behind the mutiny on the *HMS Bounty*, or rather, the screen interpretation starring the late Marlon Brando, who was so enamoured of the island that he established a hotel and had an over-water villa there.

As it is

Today, French Polynesia is a mish-mash of all these various influences. As you walk through the towns, you can see a good many locals with pronounced Caucasian features speaking either in French or the local patois.

Despite Hollywood stars such as Pierce Brosnan, Eddie Murphy and Danny DeVito winging in on private jets to spend the weekend in the elite hotels of Bora Bora, and tracts of land being bought up by the moneyed, life on the island has retained much of its original charm.

There isn't too much of a rush to seek out autographs or other unruly behaviour, as the seven policemen on this 40-sq km (approximately 15-sq mile) island are able to keep the peace.

Many Americans who visit come to seek out offspring that they may have sired and left behind, testimony to Bora Bora's role in World War II as a US supply base. Bunkers and cannons are more obvious mementoes of that troubled time.

Colonialism and its discontents

In 1885, the Polynesian islands became part of the *Établissements français d'Océanie* (EFO) [French Pacific Settlements], and in 1946 they collectively became an Overseas Territory within the French Republic.

PREVIOUS PAGES: As in days gone by, visitors journey to Bora Bora to experience its majestic beauty, warm waters and colourful island life.

OPPOSITE: Moorea is magical at sunset. Lying about 20 km (approximately 12 miles) west of Tahiti and 200 km (124 miles) southeast of Bora Bora, it is rumoured that writer James Michener based his island of Bali Hai on Moorea.

LEFT: French influence is evident in the architecture of Papeete, whose many colonial structures add a rustic air and touch of history to the town.

TOP: The origin of the custom of adorning hats with flowers is not clear, but James Morrison, second mate of the *HMS Bounty*, wrote about the visors (*traumata*) made of coconut leaves that the natives wove together in a few minutes, and which they changed several times a day. In contemporary French Polynesia, it has been modified to suit current trends.

RIGHT AND OPPOSITE BOTTOM: Tattoos serve a multitude of purposes, from markers of social status to talismans against evil, or simply as adornments for the body.

OPPOSITE TOP: Brilliant sunsets are an everyday affair on Moorea.

OPPOSITE CENTRE AND BELOW: From black pearls to humbler handicrafts, the locals have built a lucrative industry upon their creativity and imagination.

On July 22, 1957, the *EFO* was officially renamed French Polynesia. The following year, 65 per cent of Polynesians demonstrated their wish to remain linked to France in a referendum. The show of support was swiftly followed in 1961 by the construction of Faa'a international airport, which did much to open up the once-isolated islands to the rest of the world.

While there is still a strong desire to be free, economic dependence on France (as well as security and other administrative issues), makes the pursuit for total independence a less-than-enthusiastic venture. A High Commissioner represents the islands to France and oversees external matters. Since the new autonomy statute of March 2004 was introduced, however, levels of domestic and internal autonomy have risen.

Life's a beach, and then you dive in

Set in a caldera that has since been filled up with turquoise water, it's well nigh impossible to deny the urge to plunge right in.

Fortunately, there are still several places where you can observe underwater marine life up close. The fish, moray eels and manta rays that populate these waters make peaceful swimming partners, but the same can't be said of the sharks and other larger creatures that can easily go into a feeding frenzy should you get too close to them for comfort.

General rules apply when it comes to the protection of coral, which have suffered from plunder and the effects of global warming. While diving, don't stand or rest on the coral, and don't break the coral with your fins. Also, no souvenir hunting or walking is allowed on the coral reefs.

All at sea

The high temperatures and strong winds that characterise the climate make it an ideal location for anyone keen on sea sports, among the foremost being canoe racing, which is important due to its link with the original inhabitants of French Polynesia. The Hawaiki Nui Va'a, which takes place in October, attracts more than 100 teams of six rowers each, and lasts for three days. During this period, the population of Bora Bora swells from 7,000 to 11,000, doing considerable favours for the local Hinano beer industry.

The other big sport is surfing, which James Cook wrote about and which was pioneered in Polynesia. Indigenous sporting activities also include stone-lifting, which can involve boulders weighing as much as 144 kg (approximately 23 stone), and javelin-throwing, where the main targets of the sport are usually coconuts.

The islands as muses

The *tiki*, an enigmatic sculpture, often shows up on contemporary designs. However, despite the efforts of local talent, it's still the contributions of Paul Gauguin, who first came to Tahiti in 1891, that have helped to shape the images of French Polynesia in Europe. Gauguin's desire to die on these lovely islands became a reality, and his body lies in the Marquesas Islands.

The 1930s also saw the arrival of Henri Matisse and Jacques Boullaire, both of who were dazzled by the quality of light, and Boullaire especially so by the Polynesian women who became a dominant theme in his work.

A good sampling of Boullaire's work is available at the Alain & Linda Gallery. Due to the close relationship that exists between the owners and the Boullaire family, an original painting can sometimes find its way into the gallery.

A smörgåsbord of influences

On special occasions — usually a Sunday meal or wedding feast — the *four tahitien*, or open oven, dominates proceedings. This outdoor communal stove is built using volcanic stones and banana stems, upon which are placed a whole split pig, fish, bananas, tapioca and other vegetables.

Another typical dish is the *fafaru*, an appetiser made from fish that is cut into small pieces and left sealed in a bottle of sea water for a fortnight. When it's opened up, coconut milk is added and it's ready to be served. Generally in French Polynesia, breakfast and lunch are the main meals, usually comprising fish.

However, the French Polynesian diet has undergone much change over the years, due to the culinary impact of various groups of migrants. The Chinese, who are active in the commercial sphere, have had the largest influence on the diet, introducing rice, soya and noodles.

Most of the fine dining establishments in Bora Bora are located in the upper-crust hotels, of which there are 14. For anyone visiting Bora Bora, they should not leave without having tried a meal at Bloody Mary's, with its sandy floors and fresh ingredients laid out at the entrance for your selection. In this casual, laidback ambience, you can partake of fresh fare, lightly done to allow the natural flavours to come through.

Modern French Polynesian cuisine is a mix of European, Asian and traditional styles, and one can sample a taste of it at Club Med's vast buffet spreads, where the industrious kitchen staff are always ready to whip up a spread of European and local fare to tantalise the tastebuds of the international mix of customers.

A prickly matter

While the colonialists came to claim the islands of Polynesia for their own, they inevitably absorbed and brought away some of the indigenous culture and practices with them.

Tattooing, which has spread well beyond Polynesian shores, was born in this region. Tattoos in Polynesian culture are dense with symbolic meaning, and function as important markers of social status and privilege. It was the birthright of the upper classes in Tahiti, and also served as indications of sexual maturity or a means of intimidating enemies. This bodily decoration was duly frowned upon by the European missionaries seeking to spread Christianity on the islands, largely due to their connection with the pagan beliefs of the Tahitians. Associated with the deities Mata Mata Arahu and Tu Ra'i Po, who were patron divinities of the art, tattooing was reserved only for those who had undergone a long apprenticeship, and pieces of bone and a mallet were the tools of choice for the geometric designs crafted by tattooists.

Black beauties

As French Polynesia's second largest income earner, the black pearl (poe rava) is a protected jewel, encased by an oyster conforming to official regulations. In order to maintain the pearl's value, only top-grade beauties are allowed to be exported, and all come with a certification of authenticity. Contrary to popular belief, black pearls come in a myriad of colours and shapes.

While perfectly round pearls are extremely rare, other shapes produced naturally by the black-lipped *Pinctada margaritifera* species of cultivated pearl oysters, such as slightly off round and baroque, still have great value, as evidenced from the three-strand necklace of 151 natural black pearls which was sold for US$902,000 during an auction at Christie's in 1989.

It is thus little wonder the natural black pearl became known as the 'pearl of queens' and the 'queen of pearls'. It is the source of many legends, stretching all the way from China to Greece. Polynesian tales and myths about these lustrous orbs have also contributed to their mystique. One of the more famous stories has it that Oro, the Polynesian god of peace and fertility, came to earth and bestowed Te Ufi, the black-lipped pearl oyster, upon mortals.

As creations of nature, strings of these black pearls are famed for their radiance and value, and like them, the chain of islands in the Pacific, such as Bora Bora, Tahiti, Moorea and Huahine, still continue to cast their inimitable, enchanting spell upon the rest of the world.

OPPOSITE TOP, CENTRE AND BOTTOM: Nature's abundance has largely remained undisturbed on the island of Tahiti. From the pristine beaches—ideal spots for surfing or watching rainbows—to the hibiscus and other vibrant flora found at the Harrison Smith Botanical Garden, there are numerous reasons to explore the island.

ABOVE: Every corner you turn on the spectacular islands of French Polynesia, you will be met by truly dramatic, unspoiled scenery. The islands' relative isolation has done much to help preserve their natural beauty.

OPPOSITE: Island outcroppings loom large on Moorea, creating an imposing backdrop for the pleasure craft lying in the shade of leafy palm fronds.

TOP: Progress has caught up with the islands and fancy catamarans and hotels have mushroomed, attracting tourists from world over, but that has not changed the pace of life for many.

CENTRE: A regal colonial bungalow stands out dramatically against the lush greenery surrounding it.

BOTTOM: A bicycle trip is one of the best ways to enjoy the sweeping vistas and rugged scenery of Moorea.

A photographer's delight, the islands off Bora Bora offer stunning vistas. The various shades of blue, a result of the coral formations, have served as inspiration to artists as well. Paul Gauguin spent his time on these islands, painting his most famous works here. Given its reach, the surface area of Polynesia is as extensive as Europe, but with a population of just 220,000 souls.

23 bora bora, french polynesia

THIS PAGE AND OPPOSITE: With its own secluded beach, Club Med enjoys a perfect location in Bora Bora. The towering peaks behind the village provide a majestic setting, while an inviting stretch of aquamarine water lies ahead. Water sports are extremely popular, with snorkelling, scuba diving, windsurfing and sailing among the variety of options which are available for the active visitor.

LEFT: The old ways are still evident in Bora Bora. Using stones to startle the fish before catching them is an ancient tradition practiced by islanders on special occasions.

TOP AND ABOVE : Excursions from Club Med Bora Bora offer a variety of offshore options, including visits to nearby islands as well as to areas where one can feed friendly rays and sharks.

RIGHT: Modern versions of traditional craft are a common mode of transportation among the islands of French Polynesia.

MAOHI

LEFT: Shopping in Bora Bora offers a wide selection of local handicrafts as well as shops offering black pearls and art.

ABOVE: Sarongs decorated with local marine motifs make for a great wrap under the scorching sun.

OPPOSITE: Approaching Bora Bora by sea, passengers are treated to a spectacular view of the surrounding islands. Many small, uninhabited isles are perfect for a lazy day under the sun, and form emerald patches in the blue waters which are fringed with white coral.

lindeman island
australia

Australia is a land brimming with natural diversity, an asset that has been taken full advantage of by the country's tourism industry. Beyond the popular city destinations of Melbourne, Sydney, Perth and Brisbane, the wonders of nature beckon from such untouched locations as Lindeman Island, where its flora and fauna await exploration.

Lindeman Island, Australia
Whitsunday Islands, Queensland
Latitude 20° S, longitude 149° E

Paradise preserved

In terms of culture and nature, Australia has a great deal to offer. The uniqueness and range of contemporary Australian culture finds itself nestling alongside ancient aboriginal traditions, while natural attractions are carefully managed to ensure that the ecological balance in these areas is not upset by the demands of tourism.

Often called the Lucky Country, the world's largest island and smallest continent includes vineyards which produce a range of award-winning wines, emblematic fauna such as the kangaroo, koala and emu, and geographical marvels like Uluru, or Ayers Rock, the Twelve Apostles, the Great Barrier Reef and much more.

Australia also boasts a substantial number of marine attractions. There are over 7,000 beaches and close to 12,000 smaller islands in its waters. There is also a stunning variety of marine life, which includes the staggering number of over 4,000 species of fish and 150 types of sharks.

A riot of colour

The push to conserve the environment has resulted in the establishment of World Heritage Sites to ensure the continued success and viability of these areas. It's a practice that has borne fruit and in the process helped to sustain many of Australia's natural wonders, hopefully for many generations to come.

The Great Barrier Reef is a prime example of Australia's outstanding natural attractions. It has long served as a magnet for divers and those intrigued by the bustling throng of marine life beneath the waves. Stretching for 2,000 km (1,243 miles) and covering some 356,000 sq km (137,455 sq miles) in area, it was declared a UNESCO World Heritage Site in 1981. Made up of tiny polyps that accumulate in layers over each other for thousands of years, the reef is a finely balanced ecosystem that can be damaged by the slightest trace of pollution, such as human perspiration. Contrary to its name, it is not a single reef, but rather a range of approximately 2,900 individual reefs. Parts of its northern section come to within a few kilometres (approximately a mile) of the mainland, while southern sections lie hundreds of kilometres offshore.

Living polyps are responsible for the vibrant colours of the reef. When these organisms die, they produce lime, creating the hard surface which forms the substructure of the reef. It is the biggest structure in the world formed by living organisms, and the spectacular sight of the reefs underwater make it a haven for divers and snorkellers. Those less inclined to take the plunge into the waters can walk alongside stretches of reef during low tide, or view them from within the dry comfort of a glass-bottomed boat.

Whitsunday pleasures

Right in the heart of the Great Barrier Reef lie the Whitsunday Islands, a cluster of 74 islands about an hour's flight south of Cairns, in Queensland. Queensland's east coast has been a hotbed of activity since the hordes of tourists resulted in Japanese buying up large tracts of land for hotel development, and for a while the Gold Coast lived up to its name. The leisure industry in the area has seen several waves of changes, but the attraction of the sea and the myriad of gems it has to offer remains constant.

The Whitsunday Islands are fairly close together, and are truly a paradise for those who appreciate the zest and colour of marine life, an experience heightened by the clarity of the waters surrounding them. Numerous bays, beaches, and brilliant coral reefs make this a superb playground for adults and children alike.

PREVIOUS PAGES: From the air, at sea level or underwater, you are surrounded by beauty. Club Med Lindeman Island is located not far from the Great Barrier Reef, which is home to a dazzling plethora of marine life.

OPPOSITE: The aptly named Whitehaven Beach is 98 per cent pure silica. With sand good enough to make the finest optical lenses, the brilliant snowy strand stretches for almost 6 kilometres (approximately 4 miles). An exciting way to access this luxuriantly soft beach is by helicopter.

LEFT: The most developed of the seven Whitsunday islands is Hamilton Island. It is also the only island with an airport capable of handling a wide-bodied jet.

TOP: Lindeman Island's long jetty stretches out into the clear waters surrounding the island. Dolphins are frequent visitors, and if you're lucky you may even manage to spot humpback whales.

The islands are made up of national parkland which is characterised by abundant growths of rainforest, and virtually all of the islands have a reef within diving distance.

Just a number of the islands have hotels, and Club Med on Lindeman Island is an ideal spot for families. The most southerly and warmest in climate among all the Whitsunday Islands, it shares the same latitude as Hawaii.

Lindeman Island is covered by nearly 20 km (13 miles) of bushwalking tracks, and a large portion of the island is national parkland. Bird watching is excellent here, with some 90 species darting about the treetops. Blue tiger butterflies are the highlight of Butterfly Valley, and golden orchids can also be found in the mangroves.

You are also encouraged to head for Mount Oldfield, some 200 m (approximately 220 yards) above sea level, which will reward you with an exceptional panoramic vista of the Whitsunday Islands. Cameras and binoculars are essential companions if you are hoping to make the most of your trip to the peak.

For those keen on getting wet, there are seven glorious beaches, with Gap Beach being one of the best locations for snorkelling and catching oysters. Frequent visitors here include dolphins and majestic humpback whales. If you are in search of some peace and solitude, you can paddle your way to one of the tiny islands around Lindeman Island, where you can stretch and relax or live out your desert island fantasies.

An oasis of activities

Club Med Lindeman Island is located on a steep hillside overlooking some of the most tranquil islands in Kennedy Sound. The laidback , languid atmosphere which prevails in the three-storey resort is enhanced by the numerous palms which provide guests with abundant, leafy shade. In true Club Med fashion, one can find a comprehensive range of entertainment and services.

Along with the usual water sports, more adventurous visitors with a taste for the novel can join the circus school and learn the tricks of the trade, from clown acts to feats on the trapeze. Those looking for more conventional ways to pass the time can always take advantage of the putting greens set against a backdrop of radiantly azure skies and waters. Club Med Lindeman Island is also an ideal starting point from which to explore the Great Barrier Reef, and should the eager tourist in you still be looking for more, the rest of Australia awaits.

LEFT AND TOP: Lindeman Island has great views across the Whitsunday passage. You can spend an entire day in leisurely contemplation of the beautiful surroundings, or take advantage of the inviting waters.

OPPOSITE: The only resort on Lindeman Island, Club Med offers golf and nature walks in addition to a wide range of land and water activities.

OPPOSITE: Lindeman Island is 20 sq km (approximately 8 sq miles) in area, and more than 90 per cent is national parkland. Club Med occupies the entire island, which offers activities that include walking tracks which are noted for bird and butterfly spotting.

BELOW AND BOTTOM: Bird watching is one of the main outdoor activities on Lindeman Island, and you can view cockatoos, swamp hens and other indigenous birds.

RIGHT: For the golfing enthusiast, the nine-hole course set against an awe-inspiring backdrop is the ideal place in which to practice your putts.

THIS PAGE: If you've had enough of prowling the nature trails, you may want to just sit back on the beach or lie down and absorb the rays.

OPPOSITE: A dramatic work of art unfolds as nature paints a fascinating picture upon its own canvas. The shifting sands which are churned by the currents create a swirling pattern on Whitehaven Beach that is constantly evolving.

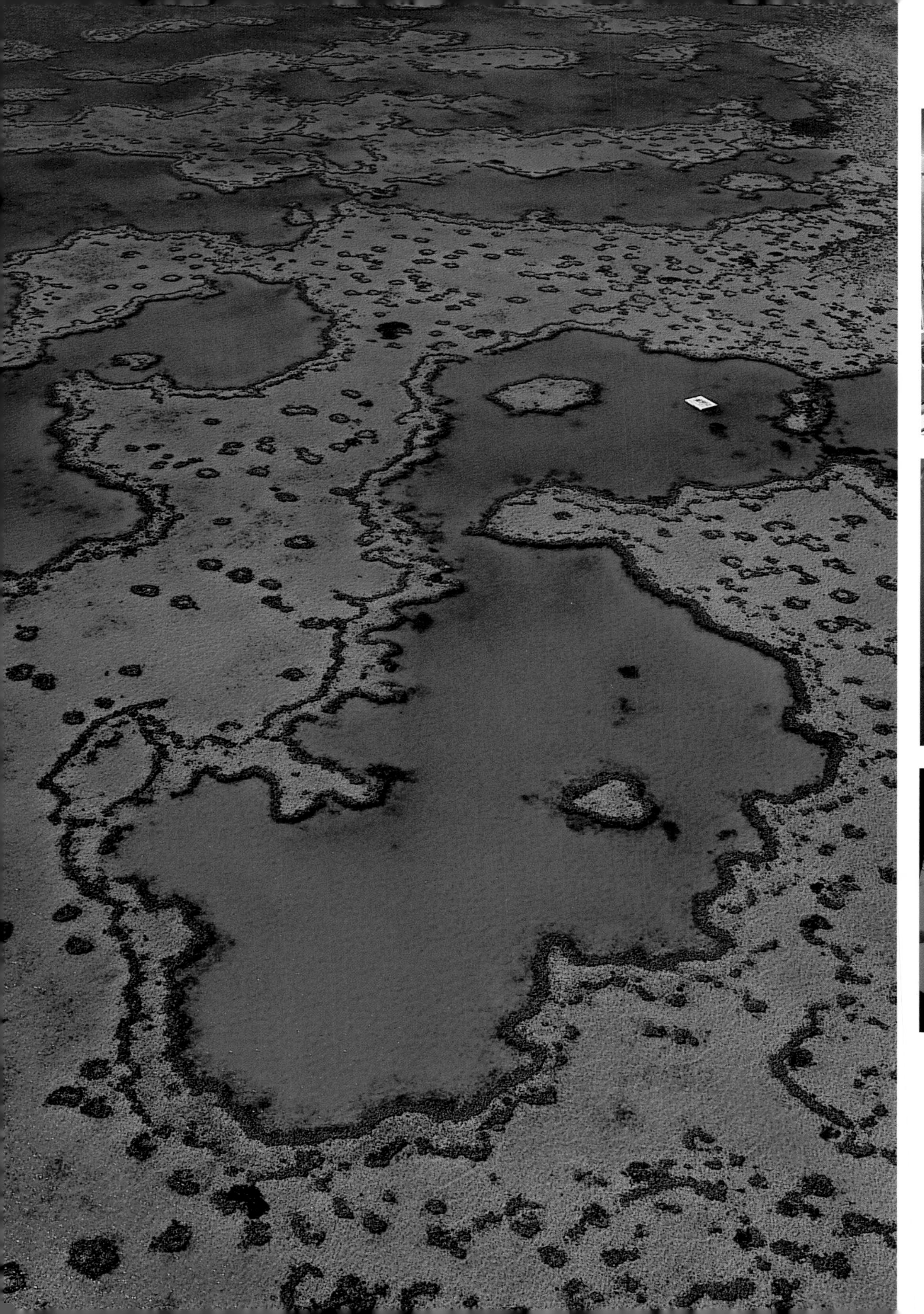

OPPOSITE LEFT: The Great Barrier Reef is a spectacular sight when viewed both from sea or air. Tucked inconspicuously within this natural wonder is a small reef shaped like a heart.

OPPOSITE TOP, CENTRE AND BOTTOM: The Reef's spendid marine life can be viewed when snorkelling or through the underwater windows of a specially outfitted dive barge.

ABOVE: Under the vigilant eye of a life-guard, snorkellers enjoy a close-up view of the abundant marine life that makes visiting the Great Barrier Reef such an unforgettable experience.

kabira
japan

The landscape of Japan has often been dominated by Tokyo, with its urban sprawl, bullet trains and an economy fuelled by a strong manufacturing sector and fast-changing trends. Contributing to the world's second-most powerful economy, however, has come at a price, and the lightning pace of life in urban Japan is such that city-dwellers feel driven to seek out places where they can rest, relax and recuperate.

Remotely resplendent

The rapid-fire pace of life and excitement in Tokyo has long played a significant role in the promotion of Japan as a tourist destination abroad. But tourism is also increasingly building up in quieter areas of the country, such as Okinawa. Okinawa used to be independent until it became a Japanese prefecture in 1879. Towards the end of World War II, it was the site of some intense fighting, serving as the last line of defence for Japan before the mainland could be breached. Almost bordering the coast of Taiwan, the Ryukyu Islands, off the mainland of Okinawa, offer a refreshing alternative to the typical images of Japan that abound in tourist guides. Here, the roof ornaments are more reminiscent of those found in China, and the swarthy Yaeyamans look

Kabira, Japan
Ryukyu Islands
Latitude 28° N, longitude 128° E

distinctly different from the average Japanese. Chinese trading vessels used to ply the waters, and some of the jetsam that washed ashore included ceramic ware and a number of other objects. A selection of these have wound up in the Shiritsu Yaeyama Museum, which offers a wealth of artefacts hinting at the cultures that have left their stamp on these islands. Ceramics, textiles, canoes and other bric-à-brac have been preserved for display and analysis.

In its quest to find a locale which was far enough from the crowds and noise of the city, Club Med opened its resort on the remote tropical archipelago of Yaeyama, about two hours by air from Tokyo. Part of Okinawa, the southernmost prefecture in Japan, this is an area of green islands, sugarcane fields and a sub-tropical climate that is ablaze with colourful blooms above ground, and a dazzling display of marine magnificence underwater. A variety of tropical and semi-tropical foliage like banyan trees make up the local flora, which are in bloom all year round. Over 3,000 species of plants dot the variegated landscape. And the deeper you journey into the mountainous regions of Iriomote and Ishigaki islands, the more exotic the plant life becomes. Underwater, Yaeyama boasts the highest number of coral species in the world, and more than a thousand types of fish. Ishigaki, which is the administrative centre of this region, is often considered Japan's final frontier, and its southernmost city. The island, located in the East China Sea, is home to almost 80 per cent of the area's population. Being somewhat distant from Okinawa and its American military bases, the Yaeyamas have been less susceptible to external cultural influences. Even World War II and the various government policies of the time seem to have passed these idyllic islands by. Ninety per cent of the islanders live in Ishigaki, whose name means 'stone wall', which the inhabitants of the island would surround their houses with to protect them from the fierce typhoons. Yaeyama was named after the 19 islands which form the archipelago, and bear a resemblance to multi-layered (*yae*) mountains (*yama*).

Islands apart

Much of the islands' success has rested on its natural bounty. The Shiraho Reef offers the world's largest example of blue coral, while Kabira Bay, on the north shore, is a quiet area with small islets. It is here that Club Med set up its resort, with the beach and small outcroppings in the water serving as a brilliant backdrop.

PREVIOUS PAGES: Facing the turquoise waters of the East China Sea, the northernmost point of Ishigaki Island is illuminated by the Hirakubozaki lighthouse. The water here drops off from 5 m (approximately 16 feet) to almost 50 metres (164 feet), and is a grèat spot for diving.

OPPOSITE: Peace and tranquillity greet visitors to the ancient Shinto shrine in the park adjacent to Kabira Bay.

LEFT: *Hoshi-zuna* are the stars of the beaches on Taketomi Island. The name translates to 'star-shaped sand', and they are formed from the remains of foraminifera, microfossils which take on this unique shape.

TOP: In the early years, there was a need to be on one's guard on these islands, which are surrounded by large expanses of water. Lookout towers like these are to be found in many spots, for keeping a watchful eye on who or what might be approaching from beyond the horizon.

Okinawa also lays claim to an illustrious cultural and artistic tradition, and Okinawan textiles are sought after throughout Japan, fetching a high price in all the major cities of the country.

Small gems

Miyako-jima is the main island in the Miyako Islands, lying about 300 km (186 miles) to the southwest of the main island of Okinawa Island. This main island is flat and almost plateau-like, and is composed of elevated coral reefs. Sugarcane fields in the area make up much of the landscape, and the island is also known for its summer triathlon. There are seven islets around the main island. Two of them, Ikema and Kurima, are connected by a long bridge, and sightseeing can best be done by car or bicycle, both of which are easily available for rent. Irihen'na-zaki Point, which projects from the northern tip of the island, gives one an excellent view of Ikema to the north, Irabu to the west, and the coral reefs below. Agarihen'na-zaki Point lies at the eastern end of the island, stretching 2 km (1 mile) long and 200 m (220 yds) wide, where unusual rock formations can be seen. Yonaha-maehama and Suna-yama Beaches are also areas worthy of a visit.

For those keen on absorbing the quiet atmosphere of Taketomi, this little island has unspoilt beaches and a great spirit of community which is rare in today's hectic world. Residents take it upon themselves to maintain their section of the public roads. A closer look at the beach reveals the *hoshi-zuna* (star-shaped sand), which are unusually shaped fossils that can be found on the western shore.

Iriomote Island's remote stretches are home to rare animal species such as the Iriomote lynx, crested serpent eagle, *semaruhakogame* turtle and the Yonaguni atlas moth, and over 90 per cent of the island is covered with semi-tropical jungle. The island is a prime spot for nature lovers, who can catch a glimpse of these fauna should they be lucky. Adding on to these natural attractions is the Iriomote-jima Onsen, Japan's southernmost hot spring, which offers you a chance to enjoy a bathing experience away from the beaches. The sea is inhabited by majestic manta rays, and spear-wielding fishermen place lights on the water to facilitate their work of catching fish and lobsters, adding a mystical glow to the dark waters lapping peacefully against the shores of the island. These quaint attractions and their serene surroundings are suspended in a rarefied world of their own, which never fail to draw the harried urbanite back for more.

OPPOSITE BOTTOM: The *habu* snake is an extremely poisonous pit viper that has finally met its match in Okinawa. The snakes are thrust into bottles of rice wine. The potent brew that results is claimed to improve back problems, cure arthritis and boost virility.

OPPOSITE TOP: Every home has its *shisa* dog to keep a watchful pair of eyes on goings-on. These gargoyle-like 'lion-dogs' are adapted from the Chinese *fu* dog and serve as guardians against evil.

OPPOSITE RIGHT: The *sanshin*, with its snake-skin body, is a popular instrument in Okinawa, and used in *minyo* performances.

ABOVE: The rolling hills near Club Med Kabira are ideal spots for a picnic overlooking breathtaking scenery and romantic sunsets.

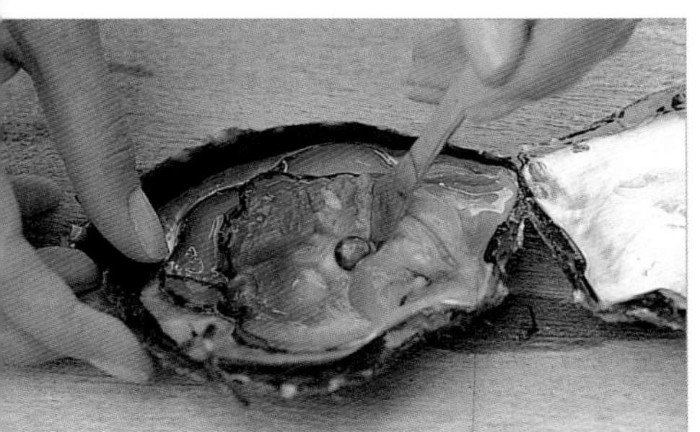

TOP: A legend telling how a merman king bestowed black pearls on a young man for rescuing his daughter was the inspiration behind this mermaid statue. In Kabira, black pearls are also believed to bring longevity and happiness to their owners.

ABOVE LEFT: The black-lip oyster is the mother shell of the black pearl, which is much sought after internationally.

ABOVE RIGHT AND RIGHT: Kabira Bay is one of two places in Japan where cultured black pearls are farmed, and its clear waters and pristine beaches are a sight to behold.

THIS PAGE AND OPPOSITE: Whether it's a romantic evening by the pool, a lazy day in the hammock, or a snorkelling trip to get up close to the manta rays, Club Med at Kabira caters to everyone. The resort offers facilities for children aged two upwards (including their own pool), while adults can enjoy an array of recreational pursuits, including various water sports and outdoor activities.

bali
indonesia

Bali possesses an innate charm, while permeated with a magical atmosphere which seems to perennially draw people to it. One of the jewels in Indonesia's crown, it has for a long time been the ultimate island escape if you were in Asia, or even further afield. Names such as Nusa Dua, Ubud and Kuta inevitably conjure up endless images of tropical bliss and a lifestyle that both envelopes and rejuvenates the body and soul.

Ambrosia for the senses

A first visit will often not be the last, and for some has proven to be a truly life-changing experience. There are many stories about foreigners who travelled to Bali out of sheer curiosity, only to be won over by a people, culture, and way of life so vastly different from theirs that they decided to sink roots on the island. Such is the allure of Bali, an island with a special enchantment and grace which beckons one towards a world that is steeped in culture and tradition.

Located in the vast Indonesian archipelago, Bali used to rely on the cultivation of rice, coffee and copra to sustain its economy. In the last few decades, however, modernisation has reached the shores of Bali, and more visitors flock to its beautiful beaches and verdant hillsides.

Bali, Indonesia
Bali Sea
Latitude 8° S, longitude 115° E

With its richly contrasting scenic attractions, the island is a photographer's paradise. From the rice paddies on the terraces of Ubud to long sandy beaches, volcanoes whose crests breach the clouds, dense tropical jungles and crashing surf, Bali's natural beauty makes it a truly memorable place. In addition to this is a people to whom hospitality is second nature, and who still hold on to the customs of their rich past in the face of the relentless surge of development.

Diversity in tradition

The origins of Bali's indigenous population date back some 4,500 years ago, when the rice-growing peoples of South China made their way to the island via the Philippines. To this day, rice still plays an important role in Balinese life. Elaborate ceremonies and festivals are centred around the rice cultivation cycle, with part of the first harvest being offered to the gods. Over many generations, the cultivation of rice has resulted in a changing landscape, with forests being cleared, hillsides terraced and an intricate network of irrigation canals put into place.

Other practices such as the chewing of betel nuts also survived the long migration process, and these customs are still practiced today in some form or fashion. Historical treasures include elaborate stone altars and stepped temples. Archaeological evidence of social stratification is evident from the various types of graves which can be found in Bali. Privileged individuals had their remains stored in sarcophagi, and the ornateness of the coffins was further proof of an individual's wealth and status.

Historical documents and early texts written in Indian languages explain some of these practices. Goa Gajah, the religious complex near the ancient capital Bedulu, is a showpiece of the Hindu and Buddhist traditions which characterise Bali itself.

Turbulent times

The island's fate was determined by its rulers' relationship with the powers of Java. In the 12th century, the ancient Majapahit kingdom of Java

PREVIOUS PAGES: The Nyepi Festival, held on the beach adjacent to Club Med Bali, celebrates the start of the Balinese new year, and is spent in quiet contemplation, a stark contrast to the noisy festivities typical of new year celebrations.

OPPOSITE: Outrigger canoes with bulging eyes and swordfish bills take you back in time, before precision-crafted catamarans invaded the waters.

TOP LEFT: Painting in the Young Artists style, pioneered by Arie Smit, one of Bali's most eminent expatriate artists.

BOTTOM LEFT: Many ateliers take in students keen on learning handicraft skills. They can often be seen sitting patiently toiling to transform a piece of wood or an empty canvas into a work of art that is up to the teacher's exacting standards.

TOP: Much of Balinese life revolves around ritual. Offerings are made here as part of the Nyepi Festival.

played an important role in shaping Balinese politics. The practice of handing down the *kris* (a lethal dagger with a waved, elaborate blade) from generation to generation, for instance, is an example of Majapahit custom, and evidence of the extent of its influence on Balinese society.

The first Europeans to set foot on Bali were the Dutch, in 1597. Their presence encouraged the slave trade with Java as a means of business, and since the island was divided into various kingdoms run by autonomous lords, the traffic of slaves in return for political favours resulted in a constant shift in the tides of domestic power.

As the Dutch began to expand their interests throughout the huge archipelago, Bali eventually became the target of various conflicts between the Dutch and Indonesians. The third such fracas eventually led to their prolongedl residency on the island, which began at the turn of the 20th century.

Any sort of political balance that was struck was rudely unhinged again by the arrival of the Japanese in 1942. Deemed the lesser of two evils, the Balinese initially attempted a strategy of cooperation, until the end of the Japanese Occupation saw the now unwelcome guests beating their hasty retreat from the island.

Eventually Sukarno, who had been brought back from political exile by the Japanese, was to play an instrumental role in keeping the Dutch from returning and galvanising the independence movement, culminating in a revolution that rolled on from 1945–49. As the first president of the Republic of Indonesia, Sukarno, who was part-Balinese, revived the image of the island as a cultural and artistic haven, and initiated the growth of its tourism industry, an area which was also given priority by the next president, Suharto.

Coupled with centuries of political upheaval, Bali has also endured a series of natural disasters. An earthquake and the eruption of Mount Batur in 1917, and the subsequent eruption of Mount Agung in 1963 have marked the island, while recent terrorist attacks have also put a big dent in the island's tourism industry. But despite such troubles, Bali's resilience has enabled it to bounce back with a renewed energy and spirit.

Attractions and distractions

When free-spirited hippies discovered the Far East in the 1970s, they were to return to their homes filled with tales of the island's beauty and mystery. Inspired by these pioneering tourists, more intrepid souls began to make the journey over, and Bali's tourism industry accelerated, buoyed by a strong sense of purpose.

The visual spectacles of the island's religious ceremonies, art and culture initially spread rapidly, and are now well-known sights the world over. Bali is also a delight for the senses. The scent of fresh fruit, the fragrance of local blooms, its spicy cuisine, and the warmth of its people can only be experienced on the island itself.

Its beaches are the key attractions, and Kuta, considered one of Asia's best strands, also boasts restaurants, boutiques and cafés which have helped make it a major draw for visitors. But other beachfront areas have also grown in prominence, and now find themselves playing host to their fair share of admirers.

Jimbaran, near Kuta, is more laidback, and is a good spot for windsurfing and sailing small craft. Sanur's fabled white sandy beaches and calm waters are an ideal place for relaxation.

Surfers and other more adventurous souls will find much to occupy them on the beaches of Uluwatu, Padang Padang and Bingin.

LEFT: Against the dramatic setting of pounding waves and swaying palms, the *kecak* dance takes place at a seaside temple in Bali.

BELOW AND BOTTOM: There are more temples per square mile in Bali than anywhere else in the world. Many objects are religious artefacts, and whether statues, rocks or trees, Balinese pay homage to places and things that are seen to be the dwelling spots of spirits.

Near Denpasar, the scenic area of Nusa Dua has devoted itself to tourism. Located here, Club Med is equipped with spa facilities to pamper its guests. It also caters to those who enjoy the outdoors, with golfing facilities, windsurfing, archery and a circus school. Families can enjoy the best of both worlds, with Petit Club Med keeping the little ones busy while their parents indulge in the range of activities available.

Those keen on diving and snorkelling should head for the eastern part of Bali. Scenic Nusa Lembongan has professional dive centres which are an excellent starting point from which you can explore the aquamarine waters. Amed's resort town atmosphere is great for snorkelling, while the premier dive site, Tulamben is just a splash away from a magnificent underwater world.

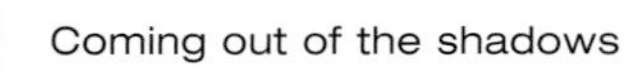

Coming out of the shadows

Balinese society observes its own practices, with every major lifestage—birth, marriage, death and others in between—marked by ritual. It is always a communal event, and the life of a typical Balinese family is intertwined with those of its neighbours.

One of the most colourful ceremonies is the cremation procession, a noisy affair that can be held a few years after the dead person has been temporarily buried. Only those of significant social standing are cremated immediately after death. As a cremation can cost a fair bit, less well-to-do families sometimes throw their lot in with another family that has a cremation planned. The body is carried alongside a bright multi-tiered tower to the cremation ground, where it is placed in a sarcophagus before the fire is lit, and is a spectacle for both locals and tourists alike.

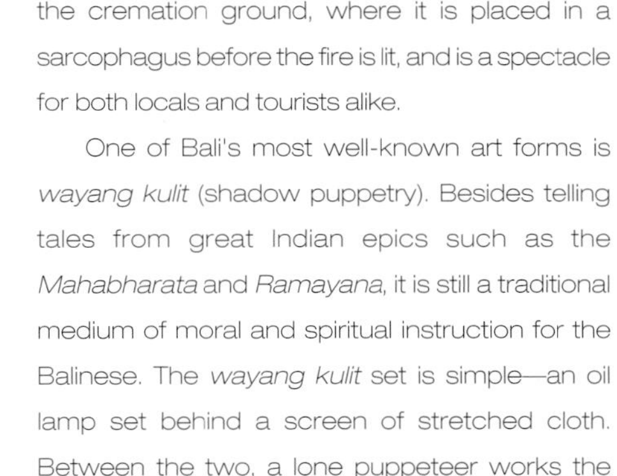

One of Bali's most well-known art forms is *wayang kulit* (shadow puppetry). Besides telling tales from great Indian epics such as the *Mahabharata* and *Ramayana*, it is still a traditional medium of moral and spiritual instruction for the Balinese. The *wayang kulit* set is simple—an oil lamp set behind a screen of stretched cloth. Between the two, a lone puppeteer works the limbs of intricately crafted puppets, bringing them to life. The puppeteer has to be highly skilled, having

to manipulate the puppets and speak their parts in a range of voices while conducting the gamelan musicians at the same time.

Music has always been an integral part of Balinese life. It is woven into the fabric of society, serving a socio-religious function. Each village has its own gamelan ensemble, whether made up of one or all 25 types of instruments, ranging from gongs and drums to xylophones. In the *kecak* dance, the voice replaces the gamelan percussion instruments. Sometimes, the *kecak* is accompanied by a lively gamelan gong orchestra in the performance of the *Ramayana*.

Dances with *topeng*, or masks, where the dancer has to adopt the character represented by the mask, are still performed in Bali, though they have now been relegated to token displays for tourists. Another dance of note is the *barong rangda* dance, in which the battle between the forces of good and evil is dramatised, and has been known to be so intense that some of the performers go into a trance. The famed *legong* dance is among the most graceful of Balinese dances, performed by elaborately dressed young girls. The male equivalent is the *baris*, or warrior dance, with war-like martial arts movements.

Cultural pursuits

After all that activity, Bali also reveals itself to be a shopper's paradise. If you like local antiques and artefacts, these are readily available, along with pottery, wrought iron and furniture. Local jewellery is also worth searching out, as some of the best designers have been trained abroad and have brought back ideas that are a fascinating fusion of East and West. Shopping hours stretch from 10 am to 10 pm, and cash is essential, with bargaining being a large part of the thrill.

Increasingly, tourists are being drawn to the stylised crafts and works which are characteristic of Balinese art. Balinese paintings have become a popular export, blending as it does Hindu and Javanese styles along with various modern influences that permeate the island's culture. The growth of Balinese art can be attributed to two European artists, Walter Spies and Rudolf Bonnet.

OPPOSITE LEFT: There are many ways of exploring Bali. The bicycle allows you to get closer to nature, and work up a sweat in the process.

OPPOSITE RIGHT: The famous rice terraces have been the source of inspiration for many artists, and a bird's-eye view of these terraces can calm frazzled nerves. Rice has been cultivated on the island for well over 2,000 years.

LEFT: One of Bali's main temples, the old Pura Ulun Danau, sits at the top of a ridge overlooking the volcano, Mount Batur.

Spies encouraged Balinese artists to bravely seek new directions. The German settled in Bali in 1927, and was instrumental in co-founding Pita Maha, the first Balinese association of artists, and was the curator of the Museum Bali in Denpasar. Bonnet arrived in Bali in 1929 and lived in Ubud, where he was happy to share his talents and knowledge with eager students. His style has strongly influenced several local artists. Another Dutch painter, Arie Smit, brought more colour to the canvas with his naïve style, which was in line with the Young Artists movement in Bali.

Other styles proliferate in Bali, key among them being the Batuan style, where every inch on the canvas is filled with intricate representations of Balinese life. If you are searching for something smaller in scale, Keliki art pieces are rarely larger than 20 cm (approximately 8 inches) by 15 cm (approximately 6 inches), and usually depict scenes of mythical characters. The Ubud style has also absorbed a considerable amount of Western influence, which can be seen clearly in the very detailed works which make up the art originating from this area. The Pengosekan style grew out of the Ubud style, and takes nature—mainly birds, insects, butterflies and plants—as its main muse and subject.

Over the years, Balinese artists have found international prominence, with Made Wianta's multi-modal works having made their mark in the installation art world. In over just two decades, this prolific and energetic artist has created over 14,000 works, which is sufficient testament to the inspiration that Bali holds for artists.

The island's skilled craftsmen have also put Bali on the artistic map with their stone and woodcarvings. Stone sculptures have been used for public buildings and temples, while the woodcarvings have been sought after by international collectors. Bali's antique furniture has also grown in repute, and is exported around the world.

Feasting in simplicity

Balinese food is a panoply of piquant spices and flavours, with rice as the staple. Meals usually consist of rice with assorted dishes of meat and vegetables, and the abundance of fresh fruit grown on the island ensures that they are constantly featured in desserts, and juices find their place on every beverage menu. With the number of foreigners on the island, however, eating options in Bali have now expanded greatly.

Bali is fast developing its culinary attractions. Apart from local fare that includes dishes such as *nasi campur* (steamed rice with spicy meat and vegetables), *satay* (grilled meat on skewers), *nasi goreng* (fried rice) and *mee goreng* (fried noodles), there are restaurants offering international fare; it's easy to find French, Italian, Japanese, Greek and other Asian cuisines.

Kuta, with its competitive environment, is home to old favourites Made's Warung and Poppies Restaurant. In Ubud, Ary's Warung offers an innovative menu, while the Dirty Duck Diner serves up its legendary crispy fried duck. Chase your meal down with a domestic beer or selection of local liquors, such as *brem* (rice wine), *arak* (local liquor) or *tuak* (coconut wine).

Nusa Dua is home to Club Med in Bali and consists of 402 rooms, making it one of the largest Club Med resorts in the Asia-Pacific and focus point for those captivated by the allure of Bali's blue waters, after which one can head inland to discover the many natural and cultural highlights of this paradise island.

LEFT AND ABOVE: Thousands of worshippers assemble on the beach near Club Med during the Nyepi Festival to pray and make offerings to the god of the sea.

RIGHT: A traditional Balinese gate such as this one at Club Med Bali is a common sight on the island, and often serves as a backdrop for cultural perfomances.

LEFT: Enthusiasts of water sports will find a wide variety of activities to keep them occupied on Lombongan Island, a 2-hour boat excursion from Bali.

TOP: White-water rafting is a popular and adrenaline-pumping activity available not far from Club Med Bali.

ABOVE: Frothing waves pound against the sheer cliffs at Ulu Watu, which is the location of one of Bali's most famous Hindu temples.

LEFT AND OPPOSITE: Club Med Bali truly encapsulates the spirit of the island. Green lawns and soaring coconut palms are matched by elements of traditional Balinese architecture which are incorporated in the design of the buildings, blending in with the island's idyllic atmosphere.

ABOVE: While the resort is built to be at one with the elements, the comfort of guests is never sacrificed. While a full range of activities is available, lounging by the pool is always an option.

bintan island
indonesia

For many tourists a holiday to Bintan is synonymous with a day trip filled with golfing and indulgence in the wide range of seafood the island has to offer. The largest island in the Riau Archipelago is a short ferry ride from Singapore, and its proximity makes it a popular holiday resort for many wanting to escape the drudgery of the workday.

Cornucopia of history

While offering all that can be expected from a tropical island resort, Bintan is also steeped in history and a rich cultural heritage that is the source of considerable charm. If you can find time to immerse yourself in the island's culture and way of life, you will find that it is filled with traditions that still hold sway in today's fast-paced world.

Delving further into the island's history, it is revealed that the Riau Archipelago had a key role to play in the region during the mid-19th century. It was the centre of the Malay world, and in 1857, scholar Raja Ali Haji from Riau wrote a book that would become the linguistic foundation for Bahasa Indonesia, Indonesia's national language.

Along with Batam, the other key island in the Riau Archipelago, Bintan has seen a lot of historical jostling for possession involving the Portuguese, Dutch and English.

Bintan Island, Indonesia
Riau Archipelago
Latitude 1° N, longitude 104° E

On another dramatic note, it was recorded that Sultan Mahmud Shah of Malacca fled to Bintan after its invasion by the Portuguese, and continued to wage battle against them from the island. The Portuguese sent an armada to Bintan in 1526 in retaliation, and soon Lingga became the capital of the Johor-Riau kingdom.

The British restored the throne to the Sultan Mahmud lineage, until a dispute over succession resulted in the kingdom being split, with territories north of Singapore falling under British rule, and the Dutch controlling those that lay to the south.

The Dutch were more intent on developing Batavia (present-day Jakarta), and as a result, nationalistic fervour developed in the neglected Riaus that sowed the seeds of resistance against the colonialists. Before becoming a province of the Republic of Indonesia, however, the island was occupied by the Japanese during World War II.

Wanderers and nomads

The roots of the Bintanese economy lie early in the 18th century, where the influence of the Bugis contributed to building a major trading centre in Bintan, attracting British, Chinese, Arab, Indian and Dutch traders alike.

Active trade and migration from China and Indochina contributed to the racial diversity that exists on the island today. The Chinese and Dutch helped develop Tanjung Pinang as a commercial town in the 1800s, while the Bugis resided on Pulau Penyengat, which lies 6 km (approximately 4 miles) away from Bintan.

For centuries before its inception as a trading centre, however, Riau was already playing host to the sea nomads of the region, the *orang laut*. Their presence is still felt in Bintan, although the forces of modernisation and development are swiftly and steadily bearing down upon them. With the pace of life on the island as a whole speeding up, the *orang laut* represent a way of life which is gradually disappearing.

Lucre and leisure

Key to Bintan's success as a trading centre is its geographical position, which undoubtedly has also played a significant role in helping it become a premier vacation spot in the region. Its proximity to Singapore also ensures that it receives a fair degree of infrastructural and economic support from the neighbouring country.

Bintan has also made the most of its location by developing holiday and golfing resorts that draw tourists from Singapore and Malaysia, with a range of options catering to various budgets.

PREVIOUS PAGES: The ocean beckons at Club Med Ria Bintan, where one can take a contemplative stroll at sunrise.

OPPOSITE: Tanjung Pinang, the main city on Bintan Island, is a mosaic of colourful sights, aromas and sounds. Water taxis make their way in between larger vessels that anchor in the harbour, a tradition that goes back to the time when it was a major trading port in the region.

LEFT TOP, LEFT BOTTOM AND TOP: Life on the island moves at an idyllic pace along colourful Trikora Beach. Stalls sell snacks and drinks, and customers can while away the time or enjoy the scenery in the shade of tall coconut palms. Fishermen bring their produce in to shore regularly, and boats can be seen undergoing repairs before the next outing.

Bright lights and city life

Tanjung Pinang, the main town of Bintan, is easily reached from Singapore. The strategic location of the harbour, close to the Straits of Malacca, makes it ideal for smaller vessels on the way from Indonesia and Singapore to cast anchor.

Once ashore, the streets are a kaleidoscope of colour and sound: a mayhem of cars, buses, bicycles and other public transport, and sometimes the sight of entire families riding on *ojek* (motorbikes), resembling a stunt carried out without the benefit of a safety net. First-time visitors will find it either a trial or thrill to navigate the roads.

Visitors who are not heading for the resorts or golf courses in the north may make their way to the nearby Sungai Ular Buddhist temple, or wander through the town and explore the sights, among them the Al-Hikmah Mosque, Vihara Bahtra Sasana Temple, and Banyan Tree Temple (named for the tree that is integral to its structure). Graveyards and monuments also serve as attractions, such as the Raja Haji Fisabillah Monument of Struggle, in memory of the hero who died in 1784 battling valiantly for Malacca against the besieging Portuguese forces.

Another tribute to an important historical figure is the restored palace of scholar Raja Ali Haji located on Pulau Penyengat, an island which lies a 15-minute boat ride away from Tanjung Pinang. The palace is a prime example of the restoration efforts that have taken place over recent years in Bintan, and highlights the island's role as the cultural capital of the Malay world in the mid-19th century. At its height, the island was home to a community of Malay literati and intelligentsia. Today, most of the 2,500 residents reel in a living based on the small fishing industry.

Piquant choices

Its location in the South China Sea means that seafood is in abundance on the island. The intense tropical heat is matched by equally spicy food, and traditional Riau cuisine includes a variety of preparation methods for mussels, fish, squid and a local shellfish, *gong gong*, a conch shell which is served with a potently spicy sauce.

Culinary staples include *otak otak*, a spicy fish snack with a mousse-like texture baked in coconut leaves, and *nasi padang*, which is made up of spicy dishes served with coconut rice.

But the options extend well beyond the local. Continental and regional fare are offered at Club Med, which aims to satisfy as many different palates as possible, given the mix of nationalities that make up the Club Med clientele.

Visitors to Bintan also should not leave without first sampling the great variety of local fruit. The durian, known locally as 'the king of fruits', is harvested in the forests south of Gunung Bintan, along with jackfruit, rambutans and dukus.

As on most other islands, the coconut tree is a mainstay of the landscape. Scenery aside, there are plantations which abound to make the most of this versatile fruit. Other plantations—tapioca, maize and peanuts—continue a tradition in agriculture that has prevailed over the years.

Inevitably, the beach will beckon, and those in search of a tan on the strand can stretch out on Trikora Beach. Popular with locals, this stretch of sandy white beach is home to a fishing village and a traditional boat-building facility nearby.

For those who would prefer more exclusive accommodation, you can find Club Med Ria Bintan at the north of the island, sitting on 500 ha (1236 acres) of parkland and offering a

LEFT, BELOW AND BOTTOM: Club Med Ria Bintan is a haven for those wishing to get away from the hustle and bustle of city life. The placid calm in the village allows you to fully unwind.

variety of activities, including Petit Club Med, Mini Club Med and Juniors' Club Med. This should leave the parents free to enjoy a sauna or to visit one of the three nearby international golf courses.

Swinging home

Golf has become a major draw in Bintan. With a number of designer courses to play on, it's not surprising that regular ferry-loads of golfers from Singapore can be seen lugging their golf caddies to make the most of their day trips.

At a fraction of the cost of playing in Singapore, Bintan boasts courses that are world-class, designed by names such as Gary Player, Greg Norman and Ian Baker-Finch.

Diving into nature

Bintan and its sister island of Batam were for a long time closed off to pleasure boats. But things have changed, and with the availability of the Indonesian government's cruising permits in Singapore, the islands in the Riau Archipelago now welcome visits by yachts and other pleasure craft.

The sea has served Bintan well. From the rich bounty of fish it offers island-dwellers to the potential for development its harbour holds, it has been a constant factor in the growth of the island.

It has also boosted tourism tremendously. Sailing and jet-skiing aside, excellent diving and snorkelling opportunities abound. From the months of April to October, the waters are calmer and allow for good visibility, and inexperienced divers can train in its relatively shallow depths. The coral reefs are easily accessible, with the chance to spot giant cuttlefish, an array of fan coral, and even the occasional dolphin.

To the northeast of the island, Berakit offers the deepest dive spot, which is suitable for the experienced diver. Those looking for some history and something a little more macabre underwater can go further east, where there are good dive spots with World War II-era wrecks to explore.

As a one-stop travel destination, Bintan packs a variety of options. Culture, history and food, as well as recreational activities and adventure are all available in equal measure on the island.

LEFT: The beaches of Club Med Ria Bintan are ideal for an early morning family outing.

ABOVE: The spa provides a wide range of treatments which helps to soothe and calm guests looking for a break from their usual hectic routines.

OPPOSITE: Sunset washes Club Med's tranquil beach in a palette of muted colours.

TOP, ABOVE AND RIGHT: A stay at Club Med is a golfer's dream. In close proximity to Club Med is the award-winning Ria Bintan Golf Club, which allows guests to enjoy a challenging game while taking in the spectacular views.

cherating
malaysia

Those in search of rural bliss amidst a setting of lush green tropical vegetation would find all that and much more in Cherating, with its simple charms and relaxed pace of life. Once a sleepy backwater town, it has grown much over the past two decades.

Variety beckons

When Club Med established itself in Cherating, it added to the choice of accommodation available to visitors, and has also added greatly to the opportunities for leisure that can be enjoyed. Tourists are now finally discovering the area's beguiling delights and diversions for themselves.

The turquoise waters off Cherating form the perfect backdrop for the next relaxing holiday on your schedule. The leisure industry has developed at a healthy pace, and the beaches of Cherating are drawing the crowds who are looking for a tranquil getaway in Southeast Asia. It is also a prime location from which to explore Malaysia, and the growth of Cherating in many ways mirrors the progress Malaysia has made in recent years.

A major business player

Situated in the heart of Southeast Asia, Malaysia is located at a crossroads for transportation by

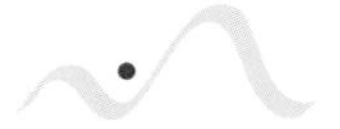

Cherating, Malaysia
Pahang
Latitude 2° N, longitude 11° E

sea or air. With Singapore to its south, it has a friendly rival for the crown of regional hub, especially with the laying of the North-South Expressway that has greatly accelerated travel between the ends of the Malaysian Peninsula.

While Singapore led the race through much of the 1980s and 1990s, the tide has started to turn somewhat. Technology has greatly helped in Malaysia's progress, and the nation has shown that it is capable of leading the way in several areas. The resource-rich country is also among the world's leading producers of commodities such as rubber, tin, palm oil, timber, pepper and petroleum. The tourism sector is also rising to become a strong revenue generator.

The Petronas Towers were the tallest man-made structures for a while, and its imposing twin structures can be seen from afar. The growth of business enterprises in the country has also helped to draw in investors keen on building upon a base of low start-up costs and inexpensive skilled labour. This is most evident in the changes taking place in Kuala Lumpur, which are being wrought at a relentless pace.

Natural charms

While Kuala Lumpur is the heart of all this activity, Malaysia has for a long time realised that its natural attractions have an appeal of their own. The magic of Malacca, Kuantan and Penang have long been chronicled by travel publications. That's just on the Malaysian Peninsula, which makes up 40 per cent of the country's land mass. Some 650 km (404 miles) across the South China Sea, the East Malaysian provinces of Sabah and Sarawak have a rugged environment which caters to the more adventurous traveller .

Malaysia's tropical climate allows for year-round tourism, though it does get very wet during the monsoon season towards the end of the year. The aboriginal islanders, or *orang asli*, made the trek from southwestern China about 10 millennia ago. As they took to the forested areas, an assortment of powerful neighbours vied for ownership of the verdant land. The Funan, Srivijaya and Majapahit empires swarmed across the land until the Chinese arrived in the city of Malacca in 1405. Together with their arrival came Islam, which was to dominate religious life in the country until the present day.

The discovery of tin and the growing rubber industry further strengthened Malaya's (as it was then known) position. As prosperity spread throughout the land, so too did word about Malacca's bounty. After the Portuguese took control of the city in 1511, the Dutch wrested it from them in 1641, followed by the British in 1975, who had been occupying Penang at the time.

In 1941, the Japanese overran the country but were beaten down largely by the brave efforts of the local guerrillas. Malaysia was formed in 1963, a confederation between Malaya, Sabah, Sarawak and Singapore, the last of which became an independent republic two years later.

PREVIOUS PAGES: Club Med Cherating Beach combines Malaysia's natural beauty with today's modern conveniences. The verdant grounds of the first Club Med in Asia are surrounded by lush tropical scenery and overlooks an expansive, breathtaking coastline.

OPPOSITE: Sunlight streaming through the trees in a rubber plantation illuminates a picturesque setting.

LEFT: Tranquillity is reflected in the lotus pads floating serenely on the waters of Tasik Chini, a popular excursion spot for guests at Club Med.

TOP: The old ways still prevail on the east coast of Malaysia, where fishing boats are moored side-by-side after a busy morning of activity.

Making progress

Today, Malaysia is a glowing example of racial harmony, with the dominant Malays living side by side with the Chinese and Indians.

While Bahasa Melayu, or Malay, is the national language, most people involved in business are conversant in English. All the ethnic groups embrace their own cultures, and some aspects, like *wayang kulit* (shadow puppetry), the martial art of *silat*, and *batik*, have become symbols of the country's adherence to tradition despite the advances of modernisation.

The beauty of Malaysia is the ease with which one can find solace from the daily grind. For the busy executive, welcome relief from the stress of the office can be found in the hills, jungles or beaches, and Pahang has an abundance of such areas. Lying on the east coast of the Peninsula, it stretches from the beaches of Cherating, and through nature reserves to the highlands that form a bony spine down the length of the country.

Rich in natural resources, Pahang's fortunes were built on its mining activities and the industry that sprang up around it. Even today, its main town, Kuantan, bears the mark of the trade that has contributed to its standing over the years.

As the administrative centre of the state, the state's ongoing development is directed from Kuantan. The growth of the industrial centre has resulted in large corporations setting up their facilities here, and once-quiet fishing villages have been galvanised into life by the arrival of oil prospectors and the tourism dollar.

Asia's first Club Med

Visitors to the state have several options. From taking a rough hike through jungle to lounging on the beaches, the lure of activity or relaxation affords you a great deal of flexibility.

Many would probably initially bypass Kuantan en route to Cherating, some 50 km (31 miles) away, where Asia's first Club Med occupies a sprawling stretch of prime beach. What used to be a sleepy village was transformed by the arrival of the resort, and the direct result of this was a vibrant mushrooming of tourist activity.

Despite the presence of Club Med, a quaint village atmosphere still lingers at Cherating. In this unique environment, traditional ways of living can still be seen as you drive along rural roads. As children scamper from wooden houses to open fields, you'll see small food stalls lining the main road serving up an aromatic buffet of hawker fare. There is also a night market on Thursdays that is packed with stalls selling all manner of attractive yet inexpensive craft and souvenir items essential for the trip home.

Day trippers

Beyond Cherating lie several day trip options. Teluk Cempedak is close to Kuantan, and is the scene of a busy beach on weekends. Here you can pick up a wide variety of local fruit, and there is a good selection of seafood restaurants.

Beserah's fishing community is well-known for its salted fish, which is a daily requirement in many households. The use of water buffalo to transport fish from the boats to the processing area is always a novel sight for the first-time visitor. The village is an excellent source of souvenirs, where one can find a *batik* factory and a selection of locally made cottage handicrafts.

Pekan is the royal capital, and arguably one of the nicest towns in Peninsular Malaysia. The ruler of Pahang resides at the Istana (Sultan's Palace), an enormous complex housing a soccer pitch, fruit farms, large buildings, and horse stables. Other attractions include the Marine Museum, State Museum and Silk Weaving Centre.

For those who fancy a spot of diving, Pulau Tioman, the largest of a group of volcanic islands dotting the South China Sea, features clear water with good reefs. Crabs can also be found in abundance at Kemaman. Other culinary delights include *pulut panggang* (grilled fish in glutinous rice); *satay*, *nasi lemak*, a popular breakfast item of rice that is cooked in fragrant coconut milk and served with anchovies, egg, peanuts and a spicy sauce; and *nasi dagang*, a glutinous rice dish cooked in coconut milk served with fish curry.

On the other side of the state, you can head to Cameron Highlands, Malaysia's prime source of vegetables and fragrant Cameronian tea. An ideal place for rambling walks and golf, you can also treat yourself to locally grown strawberries.

The other highland retreat is Fraser's Hill, where the cool mountain air, lush vegetation and nine-hole golf course will be sure to get the tense traveller into a relaxed mood.

As for Genting Highlands, you must be in the right frame of mind to visit . Developed to the hilt, this is where you can experience Las Vegas in miniature, whether gambling in style at the Casino de Genting or enjoying a wide range of dining options, unique games, horse riding, and a 19-hole golf course. They say some people never leave. All these make Club Med Cherating an ideal starting-point for those wanting to savour the gamut of flavours Pahang has to offer.

OPPOSITE LEFT: The indigenous *orang asli* community have settled along the banks of Tasik Chini, a 2-hour drive from Club Med Cherating Beach, preserving age-old traditions, including hunting with blowpipes.

OPPOSITE RIGHT AND ABOVE: Comprising 12 freshwater lakes ringed by 12,000 ha (29,643 acres) of forest, Tasik Chini is the country's second-largest natural lake. Exploring the forest makes for a day's arduous activity.

OPPOSITE TOP AND LEFT: The peaceful atmosphere at Club Med Cherating Beach ensures that one can always find a quiet spot to enjoy a good read. More boisterous activities are on hand for children, who are always welcome at Club Med.

ABOVE: The finest spa treatments are available, such as the luxurious *lulur* treatment, traditionally reserved only for brides-to-be.

RIGHT: The wide, spacious grounds welcome guests wanting to get away from the frenzied rhythms of city life.

ABOVE: Bright sails of yachts strike a colourful contrast with the azure waters off Club Med Cherating Beach.

RIGHT AND OPPOSITE: The gently sloping sandy beaches of Pahang have long drawn tourists to this region. Lined with casuarinas, coconut trees and other flora, the pristine strands and sparkling waters have been a major attraction for beach lovers the world over.

phuket
thailand

The fun never stops in Thailand. Whether you're surrounded by the bright lights and noise of Bangkok or the more laidback holiday environs of its largest island, Phuket, there's always the ceaseless ebb and flow of those who are eager for fun and adventure.

Life lived to the max

Thailand's position as a premier holiday destination was cemented early on when intrepid travellers the world over headed to this Southeast Asian country to catch the sun, sand and surf. Whether it's shopping, partying, windsurfing, island-hopping, eating or culture that you're seeking, it's all there and always accompanied with a smile.

The politeness that has always been so much a part of life and culture in Thailand has spread its reputation for hospitality far and wide, but behind this calm veneer lies an eventful history.

Wherever riches lie, prospectors will be found. And so it was with Siam, the old name for Thailand. The earliest Bronze Age civilisation on earth which dates to 3600 BC can be found on the Korat Plateau at Ban Chiang, in the northeast region of the country.

The blending of cultures was already evident in 600 BC, when first the T'ai, then the Indians,

Phuket, Thailand
Andaman Sea
Latitude 7° S, longitude 98° E

migrated to the land of Siam. They brought their culture, cuisine, language and religion to the mix that was already developing. The T'ai mapped their migratory lines according to the rivers, and settled in Burma and Vietnam.

The Khmer empire (present-day Cambodia) made its presence felt to the south of Thailand, and left behind a proud legacy of ancient stone temples and walled cities, and such temples can still be found dotted all over the country.

The Burmese made their way to Central Thailand, where they established small Buddhist kingdoms. The influence of the Mon people of Central Thailand also contributed to today's version of Thai Buddhism, and Thai script is a combination of Khmer and Mon scripts.

The Europeans were soon on the scene, beginning from 1511 when the Portuguese arrived, followed by the Dutch, English, Spanish and French. Unlike other countries in Southeast Asia that were colonised, Thailand avoided external domination both through political shrewdness and a large dash of good fortune.

Its sagacious monarchy adopted a strategy of losing small battles but winning the bigger wars to keep the would-be plunderers at bay, and this probably played a significant part in the renaming of the country in 1939 from Siam to Thailand, which means 'land of the free'.

Its success at hanging on to its sovereignty has endeared the royalty to the 60 million-strong population, who hold the present ruling family in awe and regard them as near-divinities.

Pleasure and treasure seekers

The country can be divided into four distinct regions—the mountainous lands of northern Thailand; the rainforested southern peninsula; the central plains of the Chao Phraya River; and the high northeastern plateau. Several islands have gained popularity among Europeans and Americans keen on getaways with a difference.

Some have reckoned that if you want to get all of Thailand in a nutshell, you should head straight for Phuket. It's here where you'll find the bright lights of Bangkok mirrored in the bars and club culture of Patong Beach, the glorious beaches and the excellent cuisine, along with an abundance of activities to choose from, with buses, motorcycle taxis and *tuk-tuks* making up the main modes of transport.

As Thailand's largest island, Phuket is linked to the mainland by a bridge, but has managed to hang on to its uniqueness. The combination of influences from the Chinese, Portuguese, and sea nomads who have settled in Phuket has given it a vitality that sets it apart.

Located in the Andaman Sea off the country's southwestern coast, Phuket is a great destination for those in search of thrills and adventure. Most roads radiate from Phuket Town, which lies in the southeast of the island.

PREVIOUS PAGES: Golden sunsets are to be enjoyed in their full splendour on Kata Beach immediately adjacent to Club Med in Phuket. The sea has long served as a source of nourishment for body and soul, from fishermen in long-tail boats to beach enthusiasts craving powdery strands.

OPPOSITE: Kata Beach, arguably one of the most attractive beaches on the island of Phuket, is ideal for a jog, swim or simply lounging under the sun.

LEFT: You can't complain about not having the right shoes for the occasion. Stalls crammed with all sorts of essential beachwear cater to all sizes and tastes.

TOP: The colour and noise of Patong are hard to avoid when exploring the island of Phuket. After a day spent in the solitude of a spa or on the beach, the rhythm alters as the crescendo of haggling for the best prices for souvenirs rises steadily in the evening air.

Phuket has been a favourite haunt of the sybarite, thanks to a plethora of activities in which you can fully indulge yourself. With a full spectrum of accommodation and luxurious facilities on offer, the pampering never stops. All these draw their share of the crowd, and at times the island can barely contain the throngs. Whether it's natural beauty, outdoor activities, spa treatments, the weekend night market or its food, there's no excuse not to get involved in the bustle of activity, which often carries on into the small hours.

Phuket also provides for those who prefer the quiet life. There's a lot to be enjoyed by simply sipping the refreshing juice from a young coconut or something stronger, while tucking into some freshly prepared seafood or a spicy snack from a roadside stall.

TOP AND OPPOSITE RIGHT: Phuket's first mansion cost a princely 500,000 baht in 1904, and is a landmark on the island. Now a museum, it was the home of General Tan U Yee, who arrived in the 1870s and helped develop the island.

OPPOSITE CENTRE: Shophouse doors in Phuket's historic district reveal long-standing cultural links between the island and the former Straits Settlements of Penang, Malacca and Singapore.

OPPOSITE BOTTOM: A bird's nest fern shows off its foliage, thriving in the humid tropical atmosphere of the island.

Spice of life

Thai cuisine is arguably among the finest the world has to offer. With its delicate blend of spices and contrasting flavours, both adventurous and conservative palates can be satisfied. From the flaming hot *tom yam* soup to the milder, but still spicy, curries and salads, harmony is the guiding principle. The blending of diverse tastes and flavours ensures that Thai cuisine continues to evolve in new directions to this day.

Thai chilli, lime juice, lemon grass, tamarind juice and coconut milk are among the key ingredients used to create the dishes that typify Thai cuisine. Food from each region of Thailand has its own distinctive flavour. In the south, the food tends to be sour and spicy, influenced by the Muslim preference for cumin and curry powder. In Phuket, the Chinese-Fujian influence from centuries ago has left its own unique mark.

One of Phuket's culinary must-visits is Baan Rim Pa, located along a cliff wall overlooking the water. The food is excellent, along with the attentive service, fine wines and cigars.

Another restaurant with a view is Tunka Café, which serves local fare. For a more international selection, Club Med serves a huge buffet spread featuring a variety of culinary styles

The richness of the marine life which thrives in the surrounding waters means that fresh seafood is an obvious choice. From small roadside affairs to posh restaurants, there is a wide range of seafood available, prepared in a myriad of styles. The best places, however, can usually be found with a few sound tips from the locals.

Thai desserts are a delightful way to wrap up a dinner. They range from bananas in coconut milk to sticky rice with mango, or a variety of fresh fruit such as durian, mango, custard apple and mangosteen, which are easily available.

Out there

If you like the great outdoors, a virtually endless variety of activities can be found on Phuket. The main draw on the island is scuba-diving, with various world-class sites which are accessible by a relatively short boat ride. Ranked globally by divers among the top 10, Phuket's warm waters are ideal from October to May, when the island is blessed with calm seas and relatively dry weather.

The west coast of Phuket is a diver's haven, with dive companies located along Patong, Karon and Kata beaches. If you want quieter waters away from the diving frenzy, some of the neighbouring islands are worthwhile options.

However, that doesn't apply to Phi Phi Island, where despite its congestion, the variety of marine life and coral-encrusted walls and caves makes for fascinating aquarium-like viewing.

To the south, you can find yourself swimming alongside leopard, reef and nurse sharks. Whale sharks also make the occasional appearance.

Further away from Phuket, Shark Point offers views of underwater coral to depths of up to 25 m (82 ft). Other islands where one might consider taking the literal plunge are He, Similan, Doc Mai, and Racha Noi. If you want to get a preview of what you can expect in the water, Phuket's Aquarium on Panwa Beach presents you with a dazzling display in perpetual motion.

If you want to watch turtles lay eggs or catch a sea cicada in action, Makhao Beach north of the airport has an isolated stretch of beach ideal for this. Other water-based activities include sailing, snorkelling, yachting, windsurfing and water-skiing, which the waters are excellent for.

For those who like to stretch out on the beach, there are several to choose from. Patong Beach is perhaps the most developed, with a lively collection of recreational options along its crescent-shaped bay. After a long day beneath the sun, you can retire to the shady bars which are abuzz with an assortment of nightlife activities. Karon Beach, by contrast, is the prime choice if you want some peace on a stretch of white sand.

Kata Beach is a glorious strip of sand flanked by hills and clear water, with Club Med as a key accommodation spot. The emerald island of Bu is in close proximity, and it has a village feel which changes dramatically when the annual surfing contest takes place. Another good spot for surfing is Kalim Beach, where the coral reef and tides combine to make excellent swells.

Other noteworthy spots include Naiharn Beach with its azure lagoon, while the seemingly endless strand that is Bangtao Beach belies its former incarnation as a tin mine.

Treats on land

You shouldn't restrict yourself to the beaches, for there is also much to discover inland. Rice paddies and cashew plantations are among the highlights as you sway along gently on an elephant's back. The presence of verdant natural parks helps keep the pollution levels down. Thailand has 80 national parks and 32 wildlife sanctuaries that cover almost 15 per cent of the country. Making your way through some of them is a great way to explore the bio-diversity that is as wide-ranging as the marine life of the island. The Gibbon Rehabilitation Fund, a project to return these apes to the forest, is located in Bangpae, where a small waterfall adds character to the area. Khao Phra Taeo Wildlife Park is also home to wild animals threatened with extinction. Here you'll find langurs, mousedeer, bears, monkeys and trees resonating with bird-calls. Sirinat National Park covers a large area, and also has mangroves that are home to a wide variety of rare and unusual plants.

For those who like more adventure, you can zip around in go-karts, go mountain-biking or horse-riding, or join the putters on the world-class golf courses. Set against the stunning

Andaman Sea, golf becomes a challenge for the pro as much as the amateur at Loch Palm Golf Club, Phuket Country Club, Blue Canyon Country Club and Banyan Tree Golf Course.

At night, the action takes place in the bars and cabarets. The transvestite cabaret show of Simon Cabaret has seen amazing audiences for the boys on parade, while Phuket Fantasea is reminiscent of the lights and glamour of Las Vegas, and there's no shortage of live music and bars filled with all manner of entertainment.

More conventional displays are on show in the island's museums. The Phuket Sea Shell Museum on Rawai Beach has over 2,000 species on display, while the Phuket Rare Stone Museum on Thepkasatree Road provides an esoteric experience for the curious.

For colour and spectacle, the Sunday market deserves a visit. It's a great place to pick up souvenirs on the cheap. From 'designer' jeans, glassware, compact discs and toys to electronics, you'll find them all in the chaotic, bustling environment of the market.

There are several other shopping opportunities in Phuket, ranging from local handicrafts to intricate jewellery. Traditional and contemporary ceramic pottery, and Chinese, Burmese, Khmer and Thai antiques are excellent buys. Teak, rattan and rosewood pieces also make for good purchases.

Rubies, sapphires and all manner of coloured gems are available in Phuket. If you are looking for collectibles of more intricate craftsmanship, lacquer or pewter products make good buys.

Calming down

After all that rushing about, it's time to slow things down, and Phuket has much to offer by way of opportunities for rest and relaxation.

The traditional Thai massage, *nuad bo-ram*, uses techniques which rejuvenate the body, mind and spirit, and is available at various spas on the island. If you are interested, you could even take a two-week certificate course in Thai massage at Kata Health Spa at Kata Beach. It is this irresistible combination of thrills and relaxation that draws many back to Phuket, and Club Med.

ABOVE: The long-tail boat with an outboard motor is typical of the past and present working in harmony. These boats, long since used for fishing, have been given a new lease of life, usually transporting tourists from beach to dive spot.

OPPOSITE: The various coves and bays on Phuket make it a safe haven from nature's whims. The protected beaches are splendid for water sports and other aquatic activities.

LEFT: Phang Nga Bay is noted for its limestone cliffs that tower imposingly over the iridescent waters.

TOP AND ABOVE: Nature's unfolding drama has left an awe-inspiring, unearthly landscape. A popular excursion from Phuket is a boat trip to view the cliffs and enjoy a meal at the Muslim village of Koh Panyi (TOP) and to visit nearby James Bond Island (ABOVE).

OPPOSITE, BELOW AND RIGHT: The manicured gardens of Club Med Phuket are instrumental in creating an atmosphere of calm and relaxation. While the Club Med philosophy centres around action and activity, not everything has to be done at an overly brisk pace. Wandering through its grounds or finding a quiet spot to meditate are also options well worth contemplating.

LEFT, BELOW AND OPPOSITE: On its spacious grounds, Club Med Phuket has built a sanctuary to uplift one's spirits. Traditional Thai architecture is evident in the design, which has been complemented by the provision of stress-relieving facilities such as tranquil pools one can laze in, and herbal spa treatments where guests can indulge their senses.

Club Med

faru + kani
the maldives

They appear to be almost an accident of geography—a cluster of dots closely gathered on the map in the middle of nowhere. A closer look, however, reveals them to be the prime setting for a laidback lifestyle that is centred around the beach resort and a carefree attitude to life that one is increasingly hard-pressed to find these days.

A geographical accident

Picture-postcard shots featuring coconut trees, unblemished white beaches, sparkling waters and azure skies in a tropical setting come to life on the numerous islands that make up the stunning atolls of the Maldives.

Now a haven for divers, these islands have only been welcoming visitors for the last few decades, and this growing interest has put them firmly on the tourist map.

Set in the Indian Ocean and lying south of the sub-continent, the 1,200 islands which occupy some 90,000 sq km (34,759 sq miles) of sea and land area are something of a conundrum. Not much is known about the islands' history. What is known is that they have long lain on the path of international sea-trade routes, but facts on how these islands were first inhabited are now lost, as if washed away by the endless tides.

Faru + Kani, The Maldives
Indian Ocean
Latitude 7° N, longitude 73° E

Tides of change

Archaeological finds point towards the islands being inhabited as early as 1500 BC, though little is known about the period before Aryan migrants from India settled on the islands a millennium later. They brought with them Buddhist and Hindu beliefs and practices, but the influence of Arab traders who stopped over at the islands while plying the Africa–Far East trade route has left its mark, with the majority subscribing to Islam.

Then, the islands were a natural source of two lucrative products: the cowrie shell that was used as currency in many areas of the Middle East and India, and the Maldive fish. The cowrie shell, found naturally on the beaches of the various Maldivian islands, would be exchanged for rice, spices and luxury items, a barter trade that allowed for the continued viability of these islands. The Maldive fish proved an enduring provision item for the seafarers who would pull into the islands en route to distant shores, and the intrepid Maldivian sailors who would skim across the ocean on their traditional *dhoni*. This dark-coloured, treated tuna was produced by boiling, smoking, curing and drying the fish, and was especially sought after as it was able to retain its nutritious qualities on long ship journeys.

With all the traffic, it is not surprising that the Maldives soon became a target for opportunistic visitors. The Portuguese had their anchors down for a short while before they were forced out by local troops. The lay of the islands made it difficult to monitor activities effectively, which helped the cause of the local uprising.

In a political sleight-of-hand, the Maldivians then struck a deal with the British, who were then intent on expanding their empire. By becoming a British protectorate in 1887, the defence of the islands was assured, while the management of domestic affairs was left firmly within the grasp of Maldivians. The Maldives then finally gained their independence in 1965, and became a republic three years later.

Island life

Winging into the Maldives gives you a breath-taking aerial perspective of what you can expect. As the plane heads for the airport on Hulhule, you will pass atolls, many with uninhabited islands. These islands, which average 2 m (approximately 2 yds) in height, take the form of 26 natural atolls

PREVIOUS PAGES: Jewels set in the turquoise waters, these little outcroppings are the sparkling emeralds in the necklace of islands that make up the Maldives. Many are a boat ride away, while others are a mere wade through water.

OPPOSITE: With the islands spread out over a wide area in the Indian Ocean, the seaplane is the most efficient means of transport, bringing you speedily to and fro.

LEFT: Fishing is a major occupation of the islanders, and it is always exciting to watch the catch being unloaded at Male, the capital of the Maldives..

TOP: A prominent sight as you approach Male is the Grand Mosque, with its large dome and tall minaret. Built in 1984, it is the largest mosque in the country, and accommodates more than 5,000 worshippers.

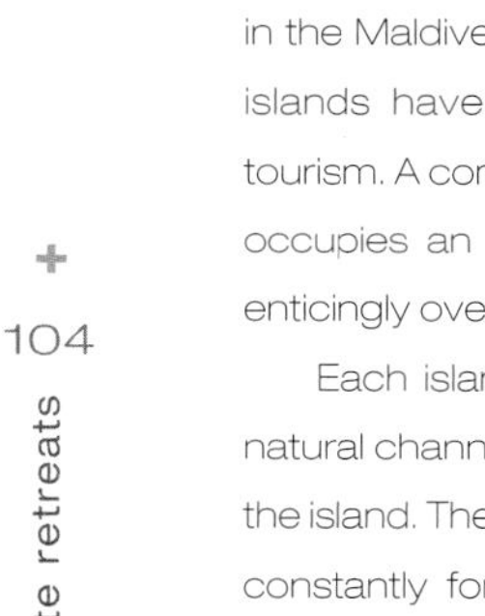

that are divided into administrative regions, also known as atolls. One gets around in water taxis and boats, which are typical modes of transport in the Maldives. Due to their small size, individual islands have been specifically developed for tourism. A common sight is the single hotel which occupies an island, with bungalows jutting out enticingly over the waves.

Each island is ringed by a coral reef, and a natural channel serves as an entry passage onto the island. The islands are actually coral which are constantly forming on the tops of submerged mountain ranges. Because they are pure coral islands, there are no discolourations due to impurities. The beaches are composed of pristine white coral sand, a truly unique feature that draws beach lovers by the droves.

Despite the large number of isles, only 200 are inhabited. Of these, Club Med has developed a couple in the North Male Atoll into villages that have drawn their fair share of enthusiastic visitors. Kanifinolhu, or Kani as many refer to it, is just a short ride from the airport by speedboat. With its tranquil lagoons and lush tropical scenery, plus the added attraction of Club Med activities, it's a haven both for those who are looking for relaxation and other more intrepid souls. The scuba base on Kani offers full-day trips, with a stop at Farukolhufushi (Faru), the other Club Med island, for lunch before returning.

Faru is surrounded by its own lagoon and coral barrier reef, with a private entrance for boats. On the island, much of the activity takes place around the surf shack and pool. For the more adventurous, there are outings for deep-sea fishing, with the catch often making an appearance on the dinner table. Excursions to nearby islands by boat or hydroplane can also be easily arranged.

At the deep end

It is what lies beneath the waves that appeals to the majority of visitors to the Maldives. One of the great dive centres in the world, the islands have near-perfect conditions for the sport. The turquoise lagoons and sheer coral drops are home to an amazing palette of colours in motion. With well over a thousand species of fish, you'll soon find yourself surrounded by groupers, surgeonfish, sharks, manta rays, angelfish, and parrotfish. The list goes on, and one has the opportunity to see all these in their natural habitat.

There are two types of dive sites specific to the Maldives. The *tilla* is a submerged pinnacle of coral that comes within a few feet of the surface. The *kandu* is a pass into the reef that, due to tidal changes, produces a lively growth of coral. With its abundance of coral species, the Maldives underwater is a confusion of activity. The best way to appreciate this is to stay in one place and take in the sights as they pass you by. A torchlight is essential for deeper dives, where the colours can astound even at those depths.

LEFT AND THIS PAGE: Various hues of blue dominate the scenery in the Maldives. At Club Med, the aim is to be in harmony with the environment. Architecture consistent with Maldivian styles can be found in the resort, but the desire for a natural ambience has created an azure paradise. Infinity pools create a sense of being at one with nature, while stunning vistas can be viewed from one of the many shady spots on its grounds.

For those who would like to bear witness to the victims of the merchant marine trade of years gone by, these beautiful islands have claimed many a vessel, like *The Corbin* (July 2, 1602), *The Hayston* (July 20, 1819) and *The Ravestein* (May 8, 1726), among others. The 82-m (270-ft) freighter, *Maldive Victory*, ran into coral in 1981 near Hulhule, taking its cargo of supplies meant for the islands to a watery grave. It has become a natural reef, and an object of curiosity for many experienced divers. Another wreck of interest is a small, inter-island freighter below 25 m (82 ft) of water, in the locale of the Ari Atoll. It has become a gathering point of sorts for the marble rays that frolic amidst its remnants.

Heading to town

While casual dress and swimsuits are the order of the day in the sweltering heat, something a little more formal is appropriate when entering the city of Male, the Maldives' political and financial hub. Working within significant space constraints, the city's planners have laid out a capital city which consists of several high-rise buildings, a main market, mosques and a maze of streets that are a pleasure to wander through.

Previously known as Sultan's Island, Male's 2 sq km (approximately 1 sq mile) holds 75,000 people. Foreign workers and tourists bring the real numbers closer to 100,000. There has been much reclamation work underway to ease the congestion, but space will prove to be a problem for quite a while to come.

The islands have only been open to tourism since 1972 due to misgivings about the ill-effects of this activity upon Maldivian society, but fear of a Pandora's box of evils being opened has been replaced by the realisation that a treasure trove of managed opportunity lies in the tourism sector.

Today, the Maldives derives the bulk of its income from three spheres—fishing, tourism and shipping. Interlinked with this is the boat-building industry, so essential to the Maldivian transport system and way of life, and handicraft-makers who ensure that tourists will leave the islands with fitting mementoes from their trip.

Many of the mosques in the Maldives are built on ruins left behind by early settlers, or Redin. These sun-worshipping people left behind beliefs and customs that are still in evidence. The imposing Huskuru Miskiiy (Friday Mosque) was built in the 17th century and is a masterpiece of the traditional art of coral curving, and the National Museum is also a worthwhile stop to make .

The sacred and the profane

One of the key landmarks on the island is the Grand Friday Mosque and Islamic Centre, with its prominent golden dome. Able to take in around 5,000 worshippers at a time, it is also an essential tourist destination and a sight to behold, especially with the sunlight glinting on its golden dome. Beyond the architectural elements, the mosque is replete with tombs of national heroes and members of royalty, giving the visitor an intriguing glimpse into the islands' past, and an insight into the religious practices of the Maldivian people.

The bustle of Male is concentrated in the local market at the northern waterfront. The fish market is an especially lively spot when the catch is brought in, and its processing of the catch is a sight in itself. The markets are also a big draw for Maldivians from the other atolls who paddle over to trade and sell their wares.

Most of the shops are along Majeedhee Magu, the main road, and they stay open till 11 pm, as browsing in the evening is a pleasant prospect after the heat of the day. Besides souvenirs, you can get canned tuna, caught by the traditional pole and line method, at a number of shops. At the north end of Chaandanee Magu is the Singapore Bazaar, named after the source of the imported wares. After a whirl through the bazaar, one should find time to rest and have a snack at one of the numerous teahouses. These boisterous cafés are plentiful, serving up a wide array of snacks and spicy meals.

Tradition on show

If it is culture that one is after, there is always the opportunity to enjoy traditional music and dance performances. The *abodu beru*, a performance involving half a dozen drummers accompanied by dancers who keep pace with the energetic rhythm, is an especially captivating spectacle. Despite its isolation, the Maldives buzzes with an alluring energy that attracts both the curious first-time tourist and seasoned traveller. Those in search of relaxation and excitement will certainly appreciate all that the Maldives has to offer.

OPPOSITE LEFT AND OPPOSITE RIGHT: Scuba divers can get a close look at the teeming rainbow of marine life in their Technicolor underwater habitat of coral gardens. One of the least exploited marine environments, the Maldives is rated among the best diving spots in the world.

ABOVE: Ringed by coral reefs, the islands of the Maldives offer shallow sandy stretches that drop off suddenly, hence the common sight of long jetties which enable access to boats and deeper water while adding to the charm of the locale.

OPPOSITE LEFT AND BOTTOM: The narrow alleyways of this island village are lost in time. Its inhabitants rely on shipbuilding and ship repairs for their livelihood, and foreign visitors rarely set foot on its shores.

OPPOSITE TOP RIGHT: Mulee Aage' was built in 1913 for the son of Sultan Shamsuddin III. However, it did not fulfil its intended purpose as the Sultan was deposed. When the Maldives became a republic in 1968 the colonial-style building built by Maldivian designers and Sri Lankan architects was designated the Presidential Palace.

ABOVE: While the *dhoni* has retained its shape and position in Maldivian society, this age-old design has proven hardy due to its practical design. These days they may be made of fibreglass, but they are still multi-purpose vessels. This particular *dhoni* is carrying sandbags to ensure the appeal of a nearby beach.

NAISH

OPPOSITE, LEFT AND BELOW: The sea and all its alluring attractions and distractions form the staple activities at Club Med. Whether you like sitting at the edge of the pier watching the fishes, running along the sand, windsurfing or para-sailing, the Indian Ocean is yours to enjoy, and Club Med provides the means to make the most of one's time there.

LEFT, TOP AND ABOVE: The word 'atoll' comes from the Maldivian language, with 26 perfect atolls making up this island chain. White beaches and waist-high water allow for endless hours of relaxation.

RIGHT: A wooden walkway stretches out into the sea, providing access to Club Med Kani's over-water bungalows.

OPPOSITE LEFT, CENTRE AND BOTTOM: Tropical bliss doesn't come any better than at Club Med Faru. Whether you're catching the setting rays from the comfort of a hammock, or getting your feet pampered at the spa, you can rest assured that you're in good hands.

OPPOSITE TOP: Swordfish is a local delicacy done to perfection in a variety of styles.

ABOVE: The poolside area at Club Med Faru is a tranquil vantage point from which to enjoy the vivid colours of the surrounding waters.

la pointe aux canonniers
mauritius

If your idea of a holiday in paradise involves long spells beneath a shady tree gazing at the horizon, sipping a long, cool drink, and alternating this inactivity with spells at a heavily laden buffet table, you should be in Mauritius, which lies in the Indian Ocean.

Fickle winds

Island getaways are now a part of many travel itineraries, but few can command the mix of exhilarating activities that Mauritius boasts. This volcanic outcropping has seen its fair share of visitors keen on claiming it for their own over the centuries. Like other gems in the Indian Ocean, Mauritius was en route to a distant land, and a convenient stop for Portuguese, Dutch, British and French trading vessels.

The influence of these erstwhile voyagers is still evident today on this small island that continues to reap the profits of its sugar trade, with textiles and tourism adding texture to the weave. The remnants of these visits are also evident in the numerous shipwrecks that can be found off Cap Malheureux (Cape of Misfortune), the northernmost tip of the island.

At one stage, sugar accounted for almost 90 per cent of exports, but in a rapidly globalised

La Pointe aux Canonniers, Mauritius
Indian Ocean
Latitude 20° S, longitude 57° E

economy, the need to keep up has resulted in diversification. Tea, fruit and tobacco have been added to the crops, while a lot of big designers stitch their clothes here. Tourism has always played an important role in Mauritius' economy. It has done much in opening up the island's economy and culture to the rest of the world, and its reputation for its coral reefs and sandy white beaches has spread far and wide.

Just so you know you're not on just another beach, you won't find yourself lying in the usual setting of a coconut tree-lined strip. Instead, you'll find casuarinas with their needle-shaped leaves and prickly fruit, and eucalyptus trees in their place, enhancing the view while acting as effective windbreaks. But there's more to Mauritius than the strand. This is also the land where the flightless dodo roamed free, until they became sitting ducks for hunters.

As you gradually make your way across the island, you'll realise how verdant it is. Mauritius is a stunning botanical display, with all manner of flora flourishing in the mixed climatic conditions. For a small island, it has contrasting weather conditions from coast to coast, depending on the wind and time of year. It's hot from January to April, with the thermometer creeping up to 35°C (95°F), along with the possibility of cyclones whipping things up. However, it's pleasant from July to September, with average temperatures ranging from 24°C (75°F) by day, and dipping to a cool 16°C (60°F) after the dramatic sunsets that fire up the sky.

Aquatic views

Surrounded as it is by water, it's only natural that this marine location is tapped to its fullest. Hotels by the water are not uncommon, with the northwest of the island being a prime spot for beach properties. Grand Baie, once a sleepy fishing village, has been transformed into the island's main holiday centre. Further along the coast, La Pointe aux Canonniers has a grand 12-km (7-mile) stretch of beach. Club Med's resort here offers great access to sea sports, and a fabulous spread that will satisfy anyone hungry after all that unbridled activity.

Off the northern tip of Mauritius, two islands, Coin de Mire and Île Plate, make for good day trips, especially if you are in search of a choice snorkelling spot. To the east of Île Plate is Îlot Gabriel, a popular lunch stop. Most hotels can also arrange for day trips to these islands.

Apart from windsurfing, kayaking, diving, snorkelling and all the other usual activities, the great opportunity to take undersea walks should not be missed. You'll feel like an aquanaut as you plod along the seabed in diving helmet and lead boots. It's also a great way for you to watch the marine world float leisurely by.

If you'd rather be above water and pulling in the big ones, deep-sea fishing opportunities abound around Grande Rivière Noire. Marlin, sailfish, barracuda, tuna and a host of other game fish are sporting possibilities. Blue and black marlin are in abundance towards the end of the

PREVIOUS PAGES: Volcanic activity in Mauritius has changed the landscape of the area. As a result of the minerals that were coughed up in the process and the chemical reactions that took place, the land's colourful hues create an almost surrealistic landscape, known as 'seven-colour earth'.

OPPOSITE: Casuarinas lined up on the beach allow you to tie your hammock to them and take it easy.

LEFT: The bracing winds blowing in from the Indian Ocean have created a more dramatic coastline in parts. Gris-Gris offers a spectacular view of nature's work.

TOP: The colonial style of Government House, built in 1738, was in honour of Mahé de La Bourdonnais, the island's most illustrious governor. His statue is close to that of Queen Victoria's, which fronts the building.

year, and attempts at tackling these aggressive creatures should only be undertaken by the brave.

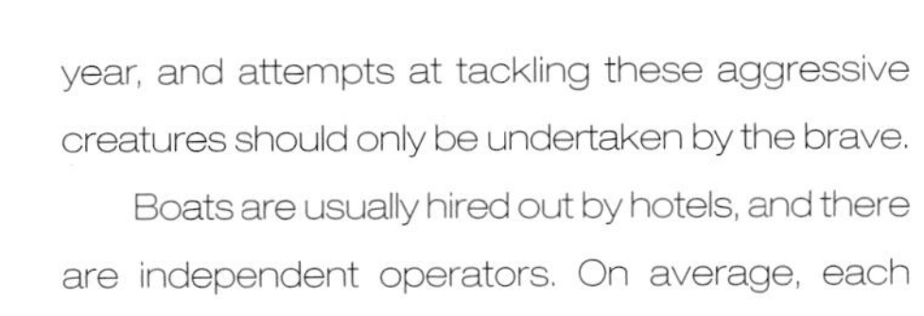

Boats are usually hired out by hotels, and there are independent operators. On average, each boat takes about six anglers, and your excursion could take from a few hours to a few days.

Like most communities that thrive on its marine bio-diversity, there are strict measures put in place to ensure its continued viability. The authorities are also very serious about enforcing rules that keep protected marine life from being smuggled out of the country.

Island rambles

Although its coastal areas are more famous, going inland can have its own charms. If you've got sturdy legs, then hiking through the forests and lowlands or bounding up some steep slopes will find you in good shape after your holiday.

To the southwest of Curepipe, nature is easily within reach as there are parklands and a mini lake district that make for gentler rambles. Most of the areas here are residential. Curepipe, a fairly prosperous market town, grew out of the 1867 malaria epidemic in Port Louis.

One of the major attractions here is the Trou aux Cerfs crater. From this perspective, you get a good view of the surrounding areas, and when you look down, you can see the heavily wooded crater of this extinct volcano.

Le Pouce, a prominent thumb-shaped peak south of Port Louis, gives you a panoramic perspective of the surrounding areas. Take a hike from Le Petrin to Grande Rivière Noire for a chance to get more intimately acquainted with the Mauritian countryside. While you are in the area, you can also appreciate the distinctiveness of Creole architecture, fine examples of which include the Curepipe Hotel de Ville, and the restored mansion, Eureka, near Moka.

Leaving it to nature

Mauritius is definitely a botanist's paradise. For the casual tourist, a visit to the Sir Seewoosagur Ramgoolam Botanical Gardens, formerly the Royal Botanic Gardens, Pamplemousses, and the Curepipe Botanical Gardens will be revealing journeys of discovery, where one can witness first-hand the bio-diversity that thrives in Mauritius. Both places are a riot of floral colour. The gardens at Pamplemousses are best seen with the help of a guide, as most plants are not labelled.

About a third of the close to 1,000 plant species on Mauritius are exclusive to the island. Apart from the extinct dodo and several of its relations, Mauritius still boasts some rare species of birds, such as the echo parakeet, Mauritius kestrel, and pink pigeon, as well as a string of assorted songbirds.

Île Ronde and Île aux Serpents are two significant nature reserves away from the main island. Île Ronde is believed to have the largest collection of endangered animal species in the world, while Île aux Serpents, devoid as it is of snakes, has been converted into a managed nature reserve. Nature lovers will also enjoy the various reserves on Mauritius, including the largest, the Black River Gorges National Park. Here you can find volcanic lakes and waterfalls, as well as unique flora and fauna.

A veritable hodge-podge

Being a volcanic island, Mauritius has an abundance of extinct craters and volcanic lakes. As you drive around the island, you'll come

OPPOSITE: While much of the fauna on Mauritius has gone the way of the famous dodo, there are still many colourful species in evidence on the island.

LEFT: The tea industry in Mauritius may not be world-renowned, but the Bois Chéri estate helps keep some of the island's traditions brewing.

CENTRE AND BOTTOM: One of the most densely populated areas in the world, the island was uninhabited until the beginning of the 17th century. Over time, the cultural and social makeup of the country has settled, with about half of the population being Hindu, and a third professing Christianity as their faith.

across neatly stacked piles of black boulders. These are lava boulders coughed up by volcanoes which were shouldered aside by labourers to make way for the many sugarcane plantations which dot the landscape. While many of these boulders have been ground down to build roads, some have been listed as monuments.

Culturally Mauritius also has much to offer, as the densely packed island of Indians, Chinese, Africans, Creoles and Franco-Mauritians works and plays together—a good example of the diversity that marks so much of Mauritian life. With over half of the population tracing their ancestry to India and professing Hindusim as their belief, the humble beginnings of the Indian community as indentured labourers belie the extent of the influence it has exerted upon the political, social and cultural landscape of Mauritius.

Most of the Hindu festivals are observed—Thaipusam where worshippers carry *kavadis*, large metal frames with spikes which are pierced through the bodies of devotees who undertake to bear them as a profession of faith; the fire-walking ceremony, Teemeedee; Deepavali, the festival of lights—along with Chinese New Year, Hari Raya Puasa, All Saints' and Christmas, reflecting the broad palette of cultures which make up Mauritius and add colour, life and depth to a holiday on the island.

Flavours to savour

The catch of the day from the waters off Mauritius is almost guaranteed to be fresh. A barbecue on the beach is often the best way to enjoy your catch. From lobsters to fish, the Creole cuisine found on the island is a happy blend of the various Indian, Chinese and European styles.

For a piquant experience that will linger on your taste buds, the Indian-influenced *carri* will offer a great taste sensation. Best savoured in a guest house or a roadside restaurant, you can also sink your teeth into fresh fish done in several styles. Snack foods are also in abundance, evidence of the national pastime of snacking. From chilli cakes to *roti*, *puri* and *samosas*, the Indian influence is again evident. Chinese dishes like pork *foo yong* and fried rice are also popular.

Port Louis

For the single-minded tourist bent on hitting the beach, they would be passing up on the chance to understand what makes this island tick if they bypassed the capital, Port Louis.

With a population of 150,000, Port Louis is the major commercial centre of Mauritius. The city is a potpourri of architectural styles, and older Creole buildings jostle for space with imposing modern highrise structures. The presence of the market at Farquhar and Queen Streets generates a lively buzz that never fails to draw in the tourists. You'll find souvenirs, traditional herbal remedies and local produce on offer. Bargaining is also a significant part of the experience.

Several museums allow you to get to know the history of this island better. The Natural History Museum & Mauritius Institute on Chaussee St features the only reconstruction of a dodo based

on a complete skeleton. The Mauritius Postal Museum near the Main Post Office has some of the most valuable stamps in the world, and the Photography Museum on Old Council St takes you back in time, with images and paraphernalia. By night, the Caudan Waterfront complex which overlooks the harbour with restaurants, bars and a casino, twinkles with life.

Song and dance

What started off as a break from the drudgery of slave labour has evolved into a dance form that is unique to Mauritius. The national dance of *sega* was usually performed around the campfire by slaves at the end of a day in the fields. Shuffling through the sand, their bodies would gyrate to the rhythm being pounded out on goatskin drums.

Today, the same hypnotic dance is still performed, not so much to relieve stress as to keep the spirit alive and to entertain the tourists. In keeping with trends, feet are still kept on the ground, but the dance has transcended its cultural context to incorporate contemporary music.

Although Western pop is now ubiquitous, a new Mauritian musical form was pioneered by the late Creole musician Kaya. By blending reggae and *sega*, seggae was born. While it hasn't spread too far beyond the shores of Mauritius, seggae has now grown to become the voice of a new generation which is eager and hungry for experimentation and innovation.

On the high seas

Once you've been to Mauritius and experienced its sense of history, it's hard not to be caught up in the spirit of things, and a model ship is a charming reminder of the island's past.

All manner of miniature ancient seagoing vessels are pieced together in accurate detail. Made out of teak or camphor wood, larger ships can take up to 400 hours to complete, and are decked with weathered sails (courtesy of a soak in some tea), making them wonderful souvenirs of your time on this remarkable island.

OPPOSITE: A day at the races in Port Louis can see crowds of up to 30,000 people. The oldest track in the southern hemisphere, Champs de Mars is also the only race course on the island.

LEFT: The colourful *sega* dances have a history that is steeped in African rhythms. Brought to the island by slaves, it is now adopted by all Mauritians.

RIGHT: With its history as a trading port, it's not surprising that the model ship-building industry has developed into a major tourist attraction and export item.

Le Morne Brabant meets the ocean at the southern tip of Mauritius. The scenery is dramatic here and the waves can be especially challenging for surfers.

125 la pointe aux cannoniers, mauritius

OPPOSITE: Nature's impact on Mauritius is evident in the stunning geographical features that have resulted, like this unusual waterfall setting.

TOP: The architectural vocabulary of Mauritius encapsulates Asian, Western and African styles.

CENTRE: Visitors can enjoy a guided tour of the Le Domaine des Aubineaux tea plantation, followed by a sip of the estate's choice pickings.

BOTTOM: The main residence at Le Domaine des Aubineaux, home of the estate's owner, was built in 1872.

THIS PAGE AND OPPOSITE: The expansive grounds of Club Med at La Pointe aux Canonniers includes lush gardens and a long stretch of beach, offering guests a wide variety of water sports. The traditional architecture and superb service capture the essence of Mauritius.

OPPOSITE: What began as a humble vegetable garden has evolved into a major tourist attraction in Mauritius. The Sir Seewoosagur Ramgoolam Botanical Gardens at Pamplemousses, named in honour of the late prime minister, includes 80 palms and about 25 species indigenous to the island. One of its impressive sights is this pond, almost obscured by giant Amazon lily pads.

TOP: Appreciating the flora and fauna sometimes requires some local knowledge and a keen eye. A weaver bird keeps vigilant watch over its intricately made nest.

CENTRE: A water lily shows off its vibrant colours.

BOTTOM: The strong winds created by the currents to the south of the island whip the island's sandy coasts.

RIGHT AND BELOW: Let the experienced hands of the Club Med masseuse knead the tension out of your body. Ayurvedic spa treatments are considered among the most beneficial for the body, and the mix of oils and herbal therapies helps to soothe tired minds and bodies.

OPPOSITE: If that isn't enough, just spend time on the lovely stretch of beach at Club Med in the company of a good book, with the knowledge that attentive service is always on hand.

Kani, The Maldives •

Faru, The Maldives •

La Pointe aux Canonniers, Mauritius •

Club Med

asia pacific / indian ocean

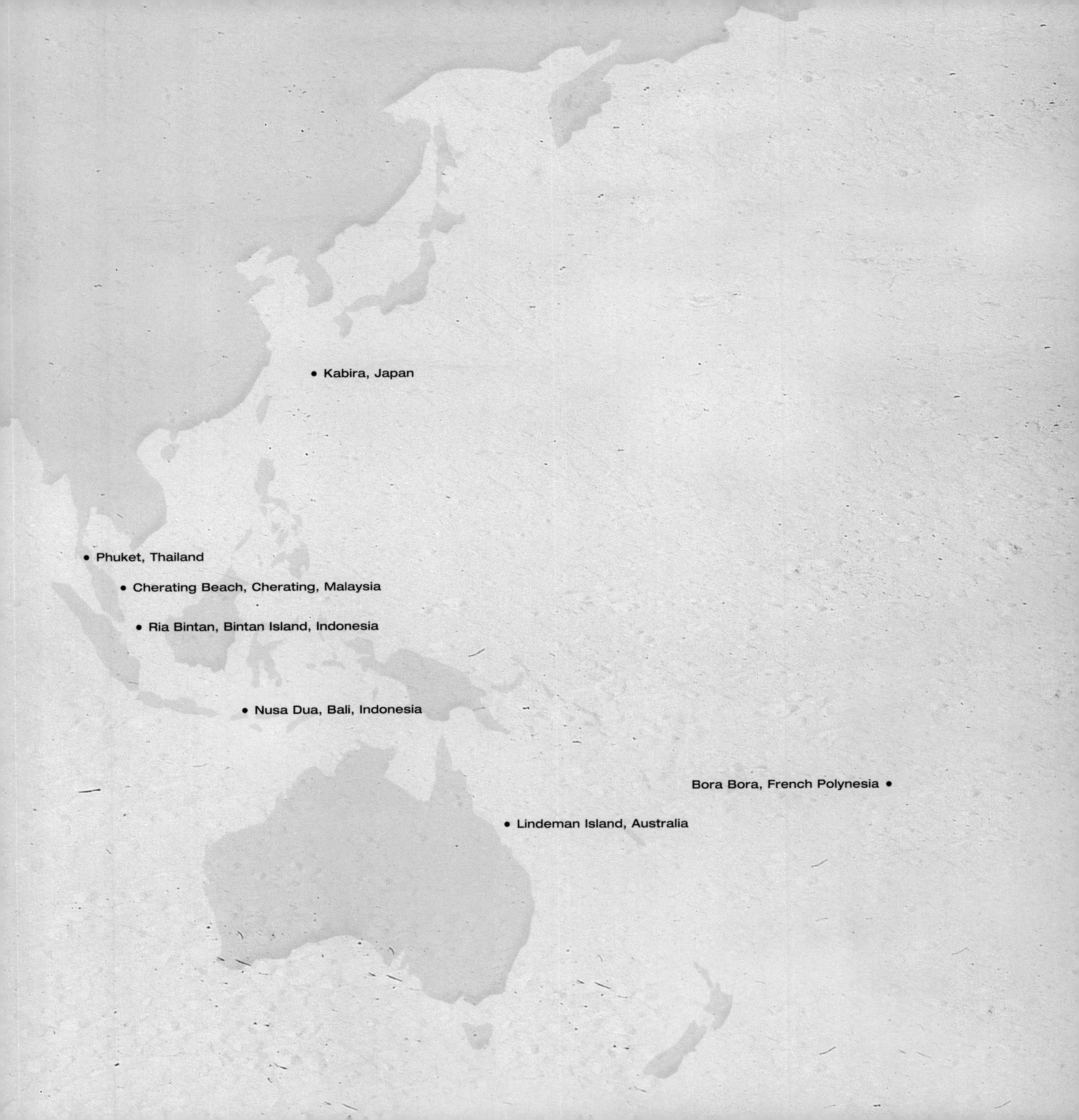

• Kabira, Japan

• Phuket, Thailand

• Cherating Beach, Cherating, Malaysia

• Ria Bintan, Bintan Island, Indonesia

• Nusa Dua, Bali, Indonesia

Bora Bora, French Polynesia •

• Lindeman Island, Australia

BORA BORA french polynesia
Club Med Bora Bora
B.P. 34 Anau
Vaitape
Bora Bora
French Polynesia
The magic of French Polynesia comes alive in breathtaking colour at Club Med Bora Bora. Surrounded by lagoons and coral beaches, guests can enjoy living in 150 Polynesian-style *fares*. Many water-based activities are available, and a wide range of refreshments is served in the restaurant and bar.

LINDEMAN ISLAND australia
Club Med Lindeman Island
P.M.B. 1, Mackay Mail Centre
Mackay
Queensland 4741
Australia
Located in national parkland on Australia's Whitsunday Islands, Club Med Lindeman Island has 218 air-conditioned rooms which are serviced by two restaurants and three bars. Activities include golf, archery, scuba diving and snorkelling, the last two of which should not be missed as the Great Barrier Reef is right at the shores of Lindeman Island.

KABIRA japan
Club Med Kabira
Ishizaki 1, Banchi
Kabira, Ishigaki City 907-0453
Okinawa
Japan
In its remote setting on a picturesque island in Japan's Yaeyama archipelago, which lies in Okinawa, Club Med Kabira brings nature right to your doorstep. Coral reefs and an azure lagoon are instant draws. For the guests staying in the 184 air-conditioned rooms, a good number of activities centre around the area's many natural attractions.

BALI indonesia
Club Med Nusa Dua
P.O. Box 7
Lot 6
Nusa Dua
Bali
Indonesia
Club Med Nusa Dua in Bali offers a stunning and tranquil haven for those in search of paradise. One of the largest Club Med resorts in the Asia-Pacific with 402 air-conditioned rooms, three restaurants and two bars, one can enjoy the natural beauty of the island as well as an array of Club Med activities.

BINTAN ISLAND indonesia
Club Med Ria Bintan
Jalan Perigi Raya, Site A11
Lago North Bintan
Bintan Utara Tanjung Uban
Indonesia
Bintan Island is the gem of the Riau Archipelago and home to Club Med's stylish village. Beaches and lush tropical vegetation serve as a backdrop to 308 air-conditioned rooms. A circus school, archery and swimming are just a few of the activities available to guests. Three restaurants and two bars provide ample entertainment during the evenings.

CHERATING malaysia
Club Med Cherating Beach
29th Mile
Jalan Kuantan Kemaman,
Cherating
26080 Kuantan
Pahang
Malaysia
Set in a remarkably clear turquoise bay, Club Med Cherating is a draw for nature lovers, occupying 80 ha (200 acres) of parkland, with surrounding tropical jungle to explore. Two restaurants and two bars service the 323 air-conditioned rooms, and guests can enjoy beach activities as well as golf, archery, flying trapeze and more.

PHUKET thailand
Club Med Phuket
3 Kata Road
Karon
A. Maung
83100 Phuket
Thailand
Beyond the hustle and bustle of Phuket City's magnetic charm, Club Med has secured a secluded spot for its property. With 307 air-conditioned rooms, two restaurants and two bars, guests can participate in the colourful activites in Phuket City, or the lively programmes offered at the village.

FARU the maldives
Club Med Faru
P.O. Box 2090
Male
The Maldives
If relaxation is something which ranks high on your list of priorities, Club Med Faru, just southwest of Sri Lanka, is the place where you can fulfil it. Picturesque scenery provides the setting for 152 air-conditioned rooms which offer comfortable accommodation. For those who fancy a bit more excitement, there is a host of sporting and leisure activities available.

KANI the maldives
Club Med Kani
Maldivian Holidays Villages Pte. Ltd.
P.O. Box 2051
Male
The Maldives
Your own private island surrounded by turquoise waters, Club Med Kani offers 209 air-conditioned rooms, two restaurants, two bars and activities ranging from snorkelling and beach volleyball to deep-sea fishing and petanque. If you're looking for a bit more bustle, the capital and main island of Male makes for an excellent day trip destination.

LA POINTE AUX CANONNIERS mauritius
Club Med La Pointe aux Canonniers
Grand Baie
Mauritius
Between the ocean and the island's charming scenery, Mauritius offers options for all manner of holidaymakers. Club Med at La Pointe aux Canonniers offers 281 air-conditioned room alongside two restaurants, three bars and leisure activities that include golf and glass-bottomed boat trips. You can also explore the island and discover its rich history and cultural diversity.

For reservations please contact:

Australia Tel: 61-2-9265 0500 • **China** (Representative) Tel: 86-21-6279 8088 • **Hong Kong** Tel: 852-3111 9388 • **Japan** Tel: 81-3-5210-6711 • **South Korea** Tel: 82-2-3452 0123 • **Malaysia** Tel: 603-2161-4599 • **New Zealand** (Representative) Tel: 0800 258 263 • **Singapore** Tel: 1800-258 2633 • **Taiwan** Tel: 866-2-2751 5511 • **Thailand** Tel: 66-2-253 0108 • website: www.clubmed.com